CONSTRUCTION SKILLS

CONSTRUCTION SKILLS

Glenn Costin

CONSTRUCTION + PLUMBING Skills Series

4e

Construction Skills
4th Edition
Glenn Costin

Portfolio manager: Sophie Kaliniecki
Product manager: Sandy Jayadev
Content developer: Margie Asmus
Project editor: Raymond Williams
Content manager: Touseen Qadri
Cover and Text Design: Cengage Creative Studio
Cover art: Antonia Pesenti - Studio Fable
Art direction: Danielle Maccarone
Editor: Sylvia Marson
Proofreader: Jade Jakovcic
Permissions/Photo researcher: Liz McShane
Indexer: KnowledgeWorks Global Ltd
Typeset by KnowledgeWorks Global Ltd

Third edition published in 2021

For product information and technology assistance,
in Australia call 1300 790 853;
in New Zealand call 0800 449 725

For permission to use material from this text or product, please email
aust.permissions@cengage.com

National Library of Australia Cataloguing-in-Publication Data
ISBN: 9780170463201
A catalogue record for this book is available from the National Library of Australia

Cengage Learning Australia
Level 5, 80 Dorcas Street
Southbank VIC 3006 Australia

For learning solutions, visit cengage.com.au

Printed in China by 1010 Printing International Limited.
1 2 3 4 5 6 7 28 27 26 25 24

BRIEF CONTENTS

CONTENTS

Guide to the text

As you read this text you will find a number of features in every chapter to enhance your study of Construction Skills and help you understand how the theory is applied in the real world.

CHAPTER OPENING FEATURES

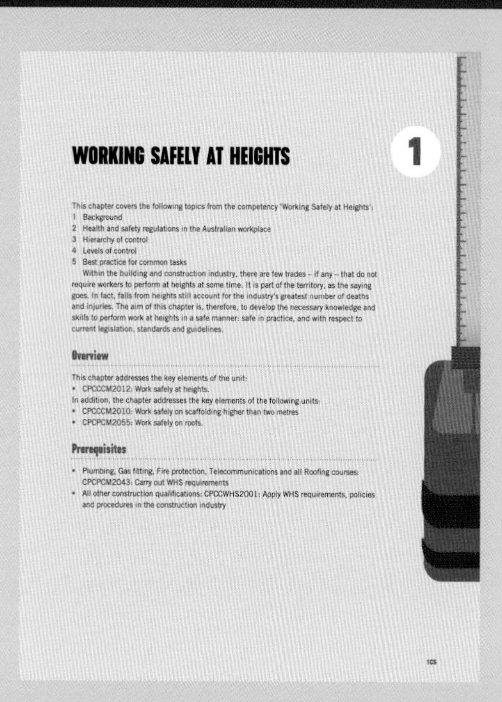

1

WORKING SAFELY AT HEIGHTS

This chapter covers the following topics from the competency 'Working Safely at Heights':

1 Background
2 Health and safety regulations in the Australian workplace
3 Hierarchy of control
4 Levels of control
5 Best practice for common tasks

Within the building and construction industry, there are few trades – if any – that do not require workers to perform at heights at some time. It is part of the territory, as the saying goes. In fact, falls from heights still account for the industry's greatest number of deaths and injuries. The aim of this chapter is, therefore, to develop the necessary knowledge and skills to perform work at heights in a safe manner: safe in practice, and with respect to current legislation, standards and guidelines.

Overview

This chapter addresses the key elements of the unit:

- CPCCCM2012: Work safely at heights.

In addition, the chapter addresses the key elements of the following units:

- CPCCCM2010: Work safely on scaffolding higher than two metres
- CPCPCM2055: Work safely on roofs.

Prerequisites

- Plumbing, Gas fitting, Fire protection, Telecommunications and all Roofing courses: CPCPCM2043: Carry out WHS requirements
- All other construction qualifications: CPCCWHS2001: Apply WHS requirements, policies and procedures in the construction industry

Identify the key concepts that the chapter will cover with the **Overview** at the start of each chapter.

FEATURES WITHIN CHAPTERS

LEARNING TASK

Learning tasks encourage you to practically apply the knowledge and skills that you have just read about.

LEARNING TASK 1.1

1 **Circle 'True' or 'False'.**
   A hazard may be defined as the likelihood of injury or harm occurring, coupled with the possible level of injury or harm that may result.
   True   False
2 **Which of the following elements of the WHS legislation is mandatory?**
   a General duty provisions
   b Detailed regulatory provisions
   c Codes of practice
   d Both 'a' and 'b'

ON-SITE

Gain insight into construction theory and how it relates to and informs day-to-day practice with the **On-site boxes**.

ON-SITE

SOLID CONSTRUCTION AND ROOFING PLUMBERS

Plumbers working on roofs should be very careful when considering aspects of an existing roof or roof elements (such as box gutters and valleys – Figure 1.12) as 'solid construction'. Box gutters are prone to rusting out, and/or their supporting members may have rotted away. In such cases, using these roof elements as walkways can lead to an exaggerated sense of safety when, in fact, a highly dangerous condition exists.

HOW TO

How to boxes provide step by step instructions on how to perform specific tasks/processes.

HOW TO

EXAMPLE

$$\begin{aligned}\text{Fall clearance} &= \text{lanyard length} + \text{energy absorber length} \\ &\quad + \text{worker height} + \text{stretch clearance} \\ &= 2.0 + 1.7 + 1.8(\text{nominal})^* + 1.0 \\ &= 6.5\ \text{m}\end{aligned}$$

*In practice, this distance is actually shorter, based upon the location of the connecting 'D' ring at the back of the harness (at least 200 mm lower, making it 1.6 m). However, we tend to err on the side of safety and leave it at 1.8 m.

KEY POINTS

Key points boxes throughout the chapters summarise the material covered in preceding sections. They are useful for quick reference and assisting with revision.

KEY POINTS

WHS (work health and safety) is not yet a nationally applicable term. OHS (occupational health and safety) still applies in Victoria.

SWA develops national WHS/OHS policy through model acts, model legislation and model codes of practice.

SWA does not enforce WHS/OHS laws – the states and territories are independently responsible.

ICONS

GREEN TIP

Avoiding the hazard of working at heights by building a platform floor on joists and then constructing framework on that platform aids doing the job efficiently which helps the environment.

Green Tip boxes highlight material that relates to environmentally sustainable workplace practices

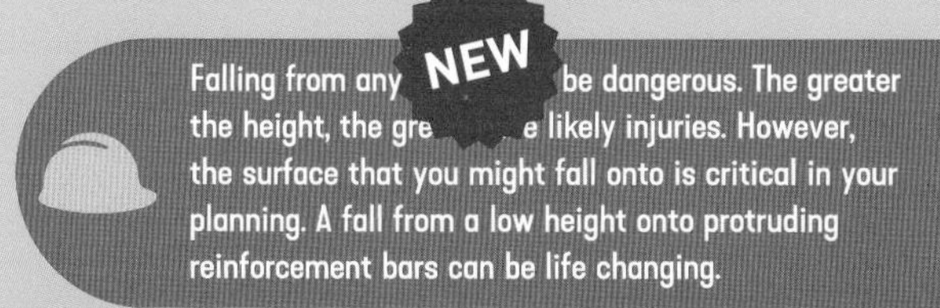

Caution boxes highlight material relating to workplace health and safety.

COMPLETE WORKSHEET 1

Worksheet icons indicate when it is appropriate to stop reading and complete a worksheet at the end of the chapter.

END-OF-CHAPTER FEATURES

At the end of each chapter you will find several tools to help you to review, practise and extend your knowledge of the key learning objectives.

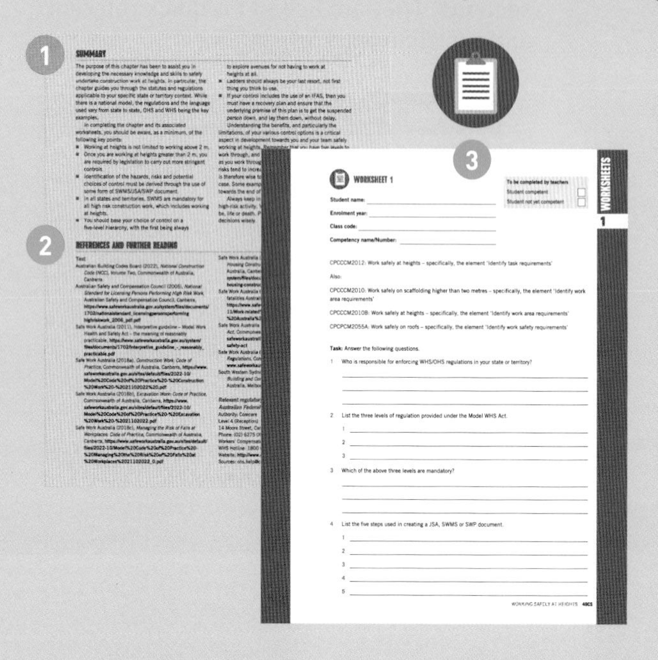

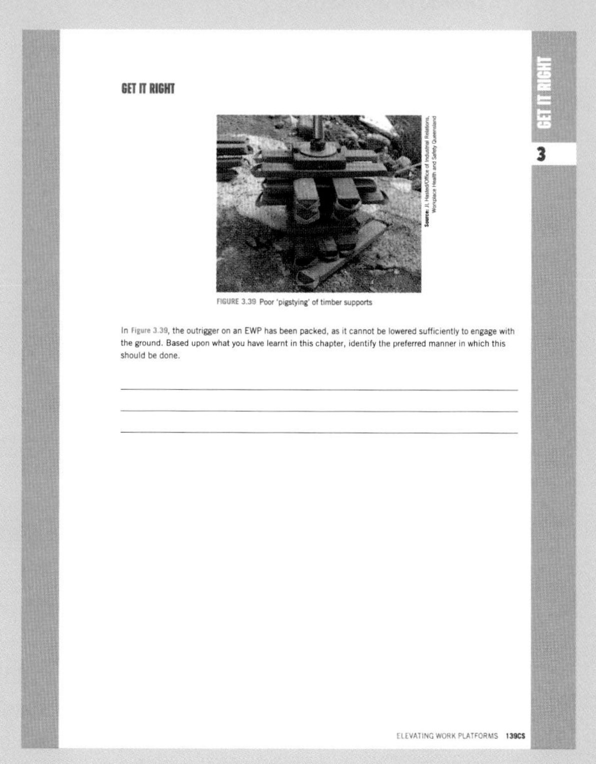

1. **Chapter summaries** highlight the important concepts covered in each chapter as well as link back to the key competencies.
2. The **References and further reading sections** provide you with a list of each chapter's references as well as links to important text and web-based resources.
3. **Worksheets** help assess your understanding of the theory and concepts in each chapter.

The Get it Right photo case study shows an incorrect technique or skill and encourages you to identify the correct method and provide reasoning.

Guide to the online resources

FOR THE INSTRUCTOR

MINDTAP

Premium online teaching and learning tools are available on the *MindTap* platform – the personalised eLearning solution.

MindTap is a flexible and easy-to-use platform that helps build student confidence and gives you a clear picture of their progress. We partner with you to ease the transition to digital – we're with you every step of the way.

MindTap for **Certificate III in Carpentry** is full of innovative resources to support critical thinking, and help your students move from memorisation to mastery! Includes:

- Construction Skills, 4e eBook.
- Basic Building and Construction Skills, 7e eBook
- Site Establishment, Footings and Framework, 5e eBook
- Advanced Building and Joinery Skills, 4e eBook
- Instructional Videos
- Worksheets
- Revision Quizzes
- And more!

MindTap is a premium purchasable eLearning tool. Contact your Cengage learning consultant to find out how *MindTap* can transform your course.

INSTRUCTOR RESOURCE PACK

Premium resources that provide additional instructor support are available for this text, including Mapping Grid, Worksheets, Testbank, PowerPoints, Solutions Manual and more. These resources save you time and are a convenient way to add more depth to your classes, covering additional content and with an exclusive selection of engaging features aligned with the text.

SOLUTIONS MANUAL

The **Solutions Manual** includes solutions to Learning Tasks, end-of-chapter Worksheets and Get It Right case studies.

POWERPOINT™ PRESENTATIONS

Use the chapter-by-chapter **PowerPoint slides** to enhance your lecture presentations and handouts by reinforcing the key principles of your subject.

ARTWORK FROM THE TEXT

Add the digital files of graphs, pictures and flow charts into your course management system, use them in student handouts, or copy them into your lecture presentations.

WORD-BASED TEST BANK

This bank of questions has been developed in conjunction with the text for creating quizzes, tests and exams for your students. Deliver these through your LMS and in your classroom.

COMPETENCY MAPPING GRID

The downloadable Competency Mapping Grid demonstrates how the text aligns to the Certificate III in Carpentry.

The Instructor Resource Pack is included for institutional adoptions of this text when certain conditions are met. The pack is available to purchase for course-level adoptions of the text or as a standalone resource. Contact your Cengage learning consultant for more information.

FOR THE STUDENT

MINDTAP

MindTap is the next-level online learning tool that helps you get better grades!

MindTap gives you the resources you need to study – all in one place and available when you need them. In the MindTap Reader, you can make notes, highlight text and even find a definition directly from the page.

If your instructor has chosen *MindTap* for your subject this semester, log in to *MindTap* to:

- Get better grades
- Save time and get organised
- Connect with your instructor and peers
- Study when and where you want, online and mobile
- Complete assessment tasks as set by your instructor.

When your instructor creates a course using *MindTap*, they will let you know your course link so you can access the content. Please purchase MindTap only when directed by your instructor. Course length is set by your instructor.

PREFACE

I have had the privilege of working and teaching in the construction industry for over 40 years. And, like most experienced tradespeople, I have worked at heights, used powder-actuated tools, installed and worked on various forms of scaffold, and set up and worked from several types of elevated work platforms (fruit picking from a 'squirrel' counts I'm sure). Despite this experience, and having taught in all four of these areas in TAFE, I would not refer to myself as 'expert'. Indeed, construction is an industry in which a misplaced belief in one's expertise can be deadly. Better that our tradespeople develop the skills and knowledge to be confident, yet cautious – aware that true competence derives from experience in multiple contexts over an extended period of time. Complacency is not an option.

Two factors encouraged me to write this book. The first was the promise (more than fulfilled) of an adequate level of technical support in reviewing each chapter. The second, and more significant, was that these four (high risk) skill areas were in need of a text that comprehensively covered all of the technical and regulatory changes that students and teachers alike need to be informed of. This fourth edition bears evidence to both the claim of relevance and the continuing technical and regulatory changes.

Why use this book? In writing the book, both Cengage and myself have acknowledged that it will be used by students and teachers alike. This being the case, we have tried to ensure that the language is appropriate, that the concepts are explored in 'trade-like' scenarios, and that the underpinning 'how' and 'why' is there to support the understanding of 'what'.

In addition, the work has been presented in a manner that is accessible and key information is easily tracked. This means that there is, at times, more information than the average student needs to know on each area, but this extra material is there to support teachers delivering the subjects. It is there to cover the 'what if' – an element critical to a firm understanding of these extremely dangerous aspects of our industry.

Too many of our young tradespeople, some in only their first few weeks of work, have died or suffered significant injury due to a lack of understanding of the risks involved in these skill areas. Between you, the teacher guiding students through these subjects, and all those involved in bringing this text into being, it is hoped that we may make a difference.

Dr Glenn P. Costin
Yackandandah, Victoria

HIGH RISK CONSTRUCTION WORK

Most of the chapters in this book deal with what is known in Australian legislation as 'high risk construction work'. High risk construction work is a very broad term and so is defined in legislation by the exact same definitive listing in each state or territory's Work Health and Safety regulations (or in Victoria, their Occupational Health and Safety regulations). The listing exampled below is drawn from the Western Australian Work Health and Safety (General) Regulations 2022.

These same regulations prescribe that *all* high risk construction work must be preceded by the development of a safe work method statement (SWMS).

Note that the very first of the legislated high risk activities (a) applies to chapters 1, 3 and 4 of this book and is highlighted by bold text. Chapter 2: Powder-actuated tools, however, is not covered by any of the activities listed below. The Model Code of Practice: Construction Work (and its state or territory equivalents, such as that of NSW) specifically discounts the activity '*(h) involves the use of explosives*' from applying to powder-actuated or explosive powered tools.

Despite this, it is this author's recommendation that you should act with caution and conduct a SWMS analysis anyway: this will include tool specific and generalised training. Justification for this stance may be found in Part 3.5 Plant (Regulation 98 (3) – Control of risk) of the Victorian OHS regulations which stipulates the use of administrative controls to reduce residual risk (Chapter 5, of all other state and territory regulations).

High risk construction work means construction work that:

a involves a risk of a person falling more than 2 metres; or
b is carried out on a telecommunication tower; or
c involves demolition of an element of a structure that is load bearing or otherwise related to the physical integrity of the structure; or
d involves, or is likely to involve, the disturbance of asbestos; or
e involves structural alterations or repairs that require temporary support to prevent collapse; or
f is carried out in or near a confined space; or
g is carried out in or near:
  i a shaft or trench with an excavated depth greater than 1.5 metres; or
  ii a tunnel;
  or
h involves the use of explosives; or
i is carried out on or near pressurised gas distribution mains or piping; or
j is carried out on or near chemical, fuel or refrigerant lines; or
k is carried out on or near energised electrical installations or services; or
l is carried out in an area that may have a contaminated or flammable atmosphere; or
m involves tilt up or precast concrete; or
n is carried out on, in or adjacent to a road, railway, shipping lane or other traffic corridor that is in use by traffic other than pedestrians; or
o is carried out in an area at a workplace in which there is any movement of powered mobile plant; or
p is carried out in an area in which there are artificial extremes of temperature; or
q is carried out in or near water or other liquid that involves a risk of drowning; or
r involves diving work.

ABOUT THE AUTHOR

Glenn Costin has more than 25 years of teaching experience in the VET sector both in Australia and internationally. In 2004 he was awarded a PhD by Royal Melbourne Institute of Technology (RMIT) for his research into skill development and its evaluation. He was also the national designer and chief judge for WorldSkills Australia for almost a decade, significantly influencing the staging of these events here and overseas. Although still actively engaged with the VET sector as both author and researcher, including consultancy and reviewing of Certificate IV and Diploma material for the Victorian TAFE Association (VTA), Glenn is currently a Senior Lecturer in Construction Technology and the Associate Head of School (Teaching and Learning) at Deakin University's School of Architecture and Built Environment in Geelong. He has also recently published another book with Cengage: *Construction Technology for Builders*.

ACKNOWLEDGEMENTS

A work of this nature requires the input of a number of technical specialists. I wish to thank and acknowledge the skills and knowledge of the text reviewers, Mark Wyborn of TAFE NSW and Jim Gott of TAFE Qld. I also wish to extend my gratitude to Honeywell Safety Products, Snorkel Australia, The GCS Group, Safe High-Ts Australia, Safe Work Australia, Ramset Australia, Buildpro Albury Wodonga, TAFE NSW South Western Sydney Institute - Miller College, and the Macarthur Building Industry Centre for their image and content contributions to the text.

Dr Glenn P. Costin

The author and Cengage Learning would also like to thank the following reviewers for their incisive and helpful feedback:

- Adam Laxton - Chisholm TAFE
- John Quine - TAFE NSW
- Kevin Tripcony - Murray Mallee Training Company
- Shane Wright - Box Hill TAFE.

COLOUR PALETTE FOR TECHNICAL DRAWINGS

Colour name	Colour	Material
Light Chrome Yellow		Cut end of sawn timber
Chrome Yellow		Timber (rough sawn), Timber stud
Cadmium Orange		Granite, Natural stones
Yellow Ochre		Fill sand, Brass, Particle board, Highly moisture resistant particle board (Particle board HMR), Timber boards
Burnt Sienna		Timber – Dressed All Round (DAR), Plywood
Vermilion Red		Copper pipe
Indian Red		Silicone sealant
Light Red		Brickwork
Cadmium Red		Roof tiles
Crimson Lake		Wall and floor tiles
Very Light Mauve		Plaster, Closed cell foam
Mauve		Marble, Fibrous plasters
Very Light Violet Cake		Fibreglass
Violet Cake		Plastic
Cerulean Blue		Insulation
Cobalt Blue		Glass, Water, Liquids
Paynes Grey		Hard plaster, Plaster board
Prussian Blue		Metal, Steel, Galvanised iron, Lead flashing
Lime Green		Fibrous cement sheets
Terra Verte		Cement render, Mortar
Olive Green		Concrete block
Emerald Green		Terrazzo and artificial stones
Hookers Green Light		Grass
Hookers Green Deep		Concrete
Raw Umber		Fill
Sepia		Earth
Van Dyke Brown		Rock, Cut stone and masonry, Hardboard
Very Light Raw Umber		Medium Density Fibreboard (MDF), Veneered MDF
Very Light Van Dyke Brown		Timber mouldings
Light Shaded Grey		Aluminium
Neutral Tint		Bituminous products, Chrome plate, Alcore
Shaded Grey		Tungsten, Tool steel, High-speed steel
Black		Polyurethane, Rubber, Carpet
White		PVC pipe, Electrical wire, Vapour barrier, Waterproof membrane

LIST OF FIGURES

WORKING SAFELY AT HEIGHTS

1

This chapter covers the following topics from the competency 'Working Safely at Heights':

1 Background
2 Health and safety regulations in the Australian workplace
3 Hierarchy of control
4 Levels of control
5 Best practice for common tasks

Within the building and construction industry, there are few trades – if any – that do not require workers to perform at heights at some time. It is part of the territory, as the saying goes. In fact, falls from heights still account for the industry's greatest number of deaths and injuries. The aim of this chapter is, therefore, to develop the necessary knowledge and skills to perform work at heights in a safe manner: safe in practice, and with respect to current legislation, standards and guidelines.

Overview

This chapter addresses the key elements of the unit:

- CPCCCM2012: Work safely at heights.

In addition, the chapter addresses the key elements of the following units:

- CPCCCM2010: Work safely on scaffolding higher than two metres
- CPCPCM2055: Work safely on roofs.

Prerequisites

- Plumbing, Gas fitting, Fire protection, Telecommunications and all Roofing courses: CPCPCM2043: Carry out WHS requirements
- All other construction qualifications: CPCCWHS2001: Apply WHS requirements, policies and procedures in the construction industry

Background

Australia is made up of six states and two territories (excluding Jervis Bay and external territories such as the Australian Antarctic Territory, Christmas Island and the like). Despite these states cooperating under a federal government, the Australian Constitution is not framed such that one single work or occupational health and safety (WHS or OHS) statute can regulate Australia as a whole.

Because of this, 10 statutes applicable to WHS or OHS remain. This cluster of statutes includes:

- two Commonwealth Acts covering federal employees and the maritime industry
- six state Acts
- two territory Acts.

To these may be added a handful of special statutes specific to the mining industry. In the past, this meant that there could be significant variation in the law between states and territories.

In 2012, a *Model Work Health and Safety Act* (Model WHS Act 2011) was adopted by most states and territories with minor variations. This was developed by Safe Work Australia (SWA), a government statutory body created to harmonise safety regulations across the nation. This meant that while the name of a state/territory WHS or OHS Act may differ, the laws are very similar.

However, not all states chose to adopt the Model WHS Act. This means that even the acronym 'WHS' does not have national application. Victoria continues to use 'occupational health and safety (OHS)' in all instances, backed by the OHS Bill passed in 2021. Western Australia adopted the term 'work health and safety (WHS)' when in the new WHS Act 2020 came into effect in March of 2022.

KEY POINTS

WHS (work health and safety) is not yet a nationally applicable term. OHS (occupational health and safety) still applies in Victoria.

SWA develops national WHS/OHS policy through model acts, model legislation and model codes of practice.

SWA does not enforce WHS/OHS laws – the states and territories are independently responsible.

Defining some terms

Before moving on, it may be useful to define the main terms that you will come across when working at heights (see Table 1.1).

See the Glossary at the end of the book for more terms and definitions.

TABLE 1.1 Key terms

Term	Definition
Administrative controls	Documents that outline a required or preferred course of action, required skills and/or supervisory tasks to be adhered to. In short, administrative controls state who, when, where and how a work activity is to be undertaken.
Anchorage point	A secure point for attaching a lanyard or other element of a travel restraint or fall-arrest system. Anchorages are designed and rated for their particular use.
Competent person	A person whose experience, knowledge, skills and qualifications provide them with the ability to perform and/or supervise a specific task or group of tasks.
Duty of care	A burden of responsibility, or legal obligation, to have thought for the safety of anyone who may be affected by your actions or, indeed, failure to act.
Elevating work platform (EWP) (also known as elevated work platform)	A mobile machine (device) that is intended to move persons, tools and material to working positions and consists of at least a work platform with controls, an extending structure and a chassis, but does not include mast-climbing work platforms.
Fall-arrest system	A system designed to stop you from falling more than a predetermined distance, and to slow you down towards the end of that distance.
Guardrailing	A protective barrier attached directly to a house, building, scaffold or other structure by posts.
Hazard	Anything, or any activity, that has the potential to cause harm. With regard to falls from a height, any situation where there is potential for someone to fall from one level to another.
Industrial rope access system	May be part of a travel restraint system or work positioning system. This line (generally steel wire rope) is used for gaining access to, and/or working at, a work face.
Inertia reel (also known as self-retracting lanyard or fall-arrest block)	A mechanical device that arrests a fall by locking up a lanyard when it is subjected to a sudden movement, much as a car seat belt does. Yet, like a car seat belt, it still allows the lanyard to travel to its full length if drawn out steadily.
Lanyard (or lanyard assembly)	A length of line (generally webbing) that connects between the harness worn by the worker, and an appropriate anchorage on the building, structure or EWP. A lanyard assembly consists of the lanyard and a personal energy absorber. The lanyard, and/or lanyard assembly, should be as short as reasonably practicable, with a working length of not more than 2 m.
Licensed	The person holds a relevant qualification from an appropriate body.

>>

Passive fall prevention device	Material or equipment, or a combination of material and equipment, that is designed to prevent a person from falling, and which, after initial installation, requires ongoing inspections to ensure its integrity, but does not require ongoing adjustment, alteration or operation to perform its function. Examples include scaffolding and perimeter guardrailing.
Person Conducting a Business or Undertaking (PCBU)	Where a business is usually an ongoing profit making organisation or system; while an undertaking may be similarly structured but is not intent on profit making.
Reasonably practicable	What could or should be done when the likelihood of risk, severity of possible injury, costs, and existing knowledge and skill are considered together.
Risk	The likelihood of an injury or some harm (such as illness) occurring coupled with the possible level of injury or harm that may result.
Safe working load (SWL)	The total weight (in kilograms) that can be on a work platform, or lifted by a cable, rope, chain or EWP. The combined weight of operators, tools, equipment and materials is not to exceed the SWL.
Scaffold	A temporary structure specifically erected to support one or more access platforms or working platforms.
Travel restraint system (also known as work positioning system)	A system incorporating a harness or belt that is attached to one or more lanyards. The lanyards are attached in turn to a static line or anchorage point. By this means, the travelling range of a worker is restricted, preventing them from reaching a position on a structure from which they could fall.

LEARNING TASK 1.1

1 **Circle 'True' or 'False'.**
A hazard may be defined as the likelihood of injury or harm occurring, coupled with the possible level of injury or harm that may result.
True False

2 **Which of the following elements of the WHS legislation is mandatory?**
a General duty provisions
b Detailed regulatory provisions
c Codes of practice
d Both 'a' and 'b'

Health and safety regulations in the Australian workplace

WHS/OHS statutes: their basic format

Irrespective of their independence, all current Australian WHS/OHS statutes, and the Model WHS Act, use a three-level approach to regulation:

- *general duty provisions* – covering everyone, from manufacturers, designers and suppliers of equipment, to employers, employees and even the self-employed. In addition, it imposes a **duty of care** upon each to people not employed, such as the general public and franchise holders – indeed, anyone who might be impacted upon by the actions of the work being undertaken
- *detailed regulatory provisions*
- *codes of practice* (compliance codes and guidance material).

Both general duties and regulatory provisions are mandatory (i.e. they must be adhered to). This means that failure to follow them can lead to criminal prosecution and/or fines.

Codes of practice (compliance codes and guidance material)

Codes of practice are a guide. (WorkSafe Victoria generally uses the term 'compliance codes', while SWA includes 'guidance material'.) Mostly, they offer advice on **hazard** identification, risk assessment and control.

Being a guide, you may choose to do other than what a **code of practice** recommends; however, should an incident occur that goes to court, the code can be used as evidence for the prosecution. In choosing not to follow a code of practice, you should make sure you have very good reasons: either by doing something that is clearly better, or because in a given work situation following the code could be dangerous (and, again, you must put in place something demonstrably better).

SWMS, JSAs, SWPs and SOPs

Currently, most OHS statutes do not rely upon the prescription of a process or steps to be followed. This is because a specific goal is often hard to identify over a broad range of activities or tasks undertaken by any one trade or group of workers. This has led to a high reliance upon documentation known as **administrative controls**. Administrative controls take the form of **safe work method statements (SWMS)** (pronounced 'swims'), **job safety analysis (JSA)** or **safe work procedures (SWPs)**. Although these latter processes still exist and have a place in developing safe work practices, it is now mandatory in all states and territories to prepare a statement (SWMS) for any high risk construction work (HRCW). Working from heights falls into this category.

A JSA or SWP document differs from a SWMS in being very task specific. They break down a task to a checklist of activities, outlining hazards and control measures. A JSA or SWP is created before a task is undertaken.

A SWMS is generally developed after a JSA or SWP, and is then adopted for frequent or regularly undertaken high risk work. A SWMS may be more detailed, but it is not a procedure like a JSA or SWP. It is a tool for confirming and monitoring required control measures.

An additional form of administrative control is a safe operating procedure, or SOP. A SOP may be one outcome of the above documentary procedures (SWMS, JSA, SWP). The purpose of the SOP is to provide a standardised way of carrying out a common task, or to use a common but dangerous tool (see **Figure 1.1**).

Source: Shutterstock.com/Chanchai phetdikhai

FIGURE 1.1 A SOP is appropriate to the use of specific equipment either on site or in a workshop setting

Across Australia, it is now mandatory for a SWMS document (specifically) to be produced prior to undertaking any HRCW. This includes risks of falling more that 2.0 m. Despite this regulatory difference, the development of a SWMS, JSA or SWP is very similar. Each covers the same basic five-step approach towards ensuring safe work.

1. *Document the activity:* With input from all involved, break up the activity into a series of step-by-step tasks.
2. *Identify the hazards:* Identify the **risks** and hazards of each task that may cause injury to the worker(s) or anyone else.
3. *Document the control measures:* Determine measures for eliminating or minimising these risks and hazards (including training or skills/qualifications required).
4. *Identify who is responsible:* Document those responsible for implementing the control measures.
5. *Monitor and review:* Monitor the activity to ensure that the documented approach to the task is being followed and that the controls are effective. Review the whole procedure on a cyclic basis (and agreed time frame, such as every six months) or when the activity changes.

The most important aspect of a SWMS/JSA/SWP document is the requirement for all workers involved in the task to have input into its development. In addition, the process must be discussed with other stakeholders – such as principal contractors, subcontractors and other workers – who might be impacted upon by the activities planned. Further, any new worker involved in the task must be made aware of, and sign on to, the documentation (JSA, SWMS or SWP).

An example of a SWMS is shown at the end of Chapter 2 (see Appendix 1 of that chapter), and further advice on how to complete one can be found in the book *Basic Building and Construction Skills* (South Western Sydney Institute of TAFE NSW, 2017).

Many of the WHS risks that impact construction work on the ground will also impact construction work at height.

Take the time to review Basic *Building and Construction Skills 6e*, Chapter 1 'Apply WHS requirements, policies and procedures in the construction industry' and review the controls that can be implemented to reduce hazards such as manual handling, working with asbestos, working with hazardous materials and working around power lines.

KEY POINTS

SWMS/JSA/SWP are all terms for administrative control documents that have a similar five-step process.

However, SWMS are legally required for all high risk construction work (HRCW) in all states and territories.

All those involved in or supervising the task should have input into preparing the SWMS. See Appendix 1 in Chapter 2 for an example SWMS document.

SWMS must describe control measures and how they are implemented, overseen and reviewed.

Enforcement

For further information on codes of practice, regulations, acts and other documentation, access and review *Basic Building and Construction Skills 6e*, Chapter 1 'Apply WHS requirements, policies and procedures in the construction industry' and Chapter 4 'Carry out workplace communication' for information on how to access these documents.

Each of the state or territory statutes provides for an overarching information, advisory and enforcement body that both develops WHS/OHS strategies and ensures their fulfilment. In addition, the statutes provide for inspectors from these bodies with the power to issue notices of improvement or prohibition, and/or to initiate prosecutions when breaches have been found. The main deterrent open to inspectors is the issuing of a fine; however, criminal proceedings can arise from failure to comply with WHS/OHS legislation, particularly when an injury or death has occurred.

Inspectors don't make an appearance just when there is a complaint or an incident has occurred. Random audits and inspections have been shown to be particularly effective, both as a deterrent and in exposing those attempting to flout the regulations.

All state and territory statutes provide for worker-elected WHS/OHS representatives. These representatives have the authority to inspect and, where necessary, issue provisional improvement notices or order that work practices identified as unsafe be discontinued.

You and WHS/OHS

Workers have an important part to play in WHS/OHS, both generally and specifically. As mentioned previously, workers can be elected as WHS/OHS representatives within a company or organisation, and they have significant powers and influence. In addition, workers must take a proactive approach to WHS/OHS (i.e. they must take action before something happens). This is done using SWMS or similar documentation. Likewise, they have a duty of care not only to their fellow workers, but to anyone who might be impacted upon by the work being done. Your role is to think about safety constantly, then to act with this awareness so as to prevent accidents before they happen: in WHS/OHS, as with most things, foresight is far more valuable than hindsight.

PCBUs

A person conducting a business or undertaking (PCBU) has prescribed duties under the WHS/OHS legislation of all states and territories except Victoria. Victoria does not recognise the term PCBU, preferring to use 'Employer' instead.

A PCBU's primary duty is to the health and safety of workers while at work and that of others who may be affected by such work. This may be said equally of Victorian employers as well. Known as their Primary Duty of Care, it is a key component of all WHS legislation. Although the term does not exist in Victorian OHS legislation, it is inferred by way of 'applicable duty', 'conduct' and 'negligence'.

The concept of 'applicable duty' or 'duty of care' is relevant to all workers. Which includes you! You are owed a duty of care by your employer, but also by all the workers around you, including other apprentices or trainees. You owe it to your co-workers, including your employer. Negligent conduct can go both ways – so we must all look out for each other, constantly. Hence catch phrases such as: 'Think Safe, Work Safe, Be Safe!'.

WHS/OHS statutes and working at heights

Working at heights is an area where WHS/OHS statutes and regulations have begun to show some significant success. In 2014, the fatalities caused by falls from heights accounted for 58 per cent of all deaths in the construction industry for that year. Statistics for 2020 suggest that the current five-year average has been reduced to 31 per cent (SWA, 2021, p. 21). Vigilance in this area remains paramount, as even these figures are too high. The key areas of concern are indicated by these fatalities statistics:

- 29 per cent from falls from buildings and structures (excluding scaffolding)
- 21 per cent from ladders.

Particularly concerning is the age breakdown:

- 27 per cent of deaths are people aged under 34
- 62 per cent of deaths are people aged over 45 (SWA, 2021, pp. 19–21).

These figures suggest that within the industry, younger workers are heeding the call for vigilance. However, older workers continue to try and get away with poor work practices. Know now that *it can*, and *it will*, if you don't think your work practice through and take into account the worst-case scenario for a particular activity. In many cases when working at heights, that means death, or at least lifelong disability.

KEY POINTS

The guidance offered to you in this text, by experienced co-workers, and/or through the WHS/OHS statutes or codes is offered for a reason. It is intended to help save your life and the lives of others. Follow it, or do something that experience has shown is better, but **do not** ignore it.

Specific WHS/OHS regulations

It is not within the scope of this text to cover specific regulations from all state and territory WHS/OHS statutes with respect to working at heights. Instead, given the shift in 2012 to the Model WHS Act and Regulations (both updated in 2022: see SWA, 2022a and 2022b), we will consider that statute alone.

The overarching emphasis of these documents is summed up in regulation 297, which requires that:

> A person conducting a business or undertaking must manage risks associated with the carrying out of construction work in accordance with Part 3.1 (SWA, 2022b, p. 212).

The emphasis on falls particularly, however, comes from Part 4.4 of the document and regulations 78 to 80. This section spans four pages and includes multiple regulations covering all aspects of fall prevention.

The best document that encapsulates the intent of these regulations is the 2018 edition of the code of practice *Managing the Risk of Falls in Housing Construction*, produced by SWA (2018d). This may be downloaded from the internet (see 'References and further reading' at the end of this chapter) and is the model for similar state and territory codes of practice throughout Australia. It is this code that shall be referred to throughout the rest of this chapter: we will call it simply the 'National Falls COP'.

Working at heights: what does this really mean?

For those new to the industry, working at heights might imply that you are working 'a long way up'. In this you would be mistaken. In fact, most statutes use anything above a mere 2 m as the point of focus, with the National Falls COP requiring risk assessments and administrative controls as a minimum for work where you might fall less than 2 m.

Put simply, anytime you step off the ground or floor level, or are near an edge or trench of any depth, you should have a SWMS or the like in place. This includes the use of common, yet specialised, equipment designed for the construction industry (i.e. rated to hold 120 kg or greater), such as:

- plasterer's stilts (Figure 1.2)
- step ladders
- stools or trestles (Figure 1.3)
- ladder bracket scaffolds.

Where the fall can be 2 m or greater, however, more stringent conditions apply.

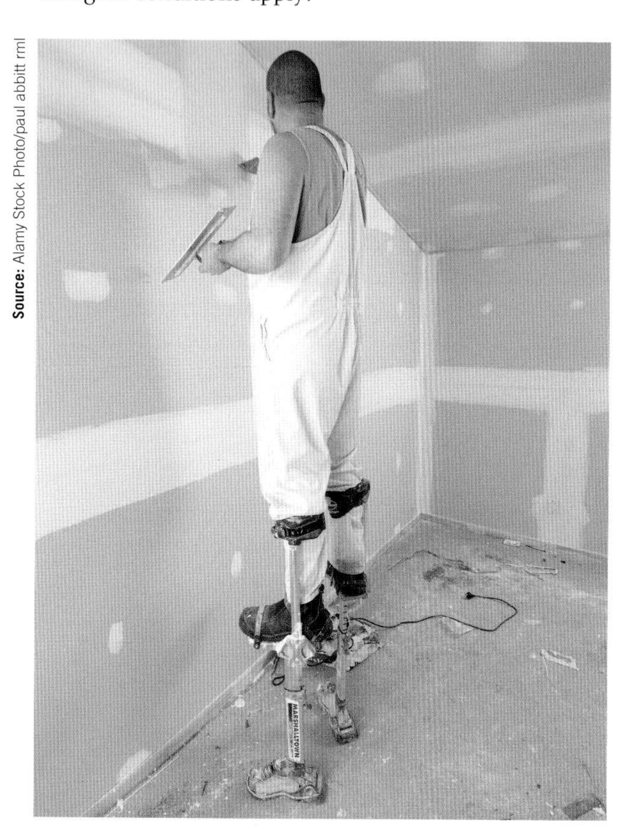

Source: Alamy Stock Photo/paul abbitt rml

FIGURE 1.2 Plasterer's stilts

Source: Alamy Stock Photo/Paul Mcguire

FIGURE 1.3 A painter's or plasterer's stool (also known as a trestle)

KEY POINTS

- Anytime you step up off floor level, you are working at heights.
- 2 m is the height at which specific regulations come into force.

Falls greater than 2 metres

Figure 1.4 provides some clear examples of the potential falls that workers face in just the domestic construction industry. For this reason, when working at heights where you or others might fall further than 2 m, your first consideration is: 'Do I need to be up there at all?' This is the beginning point of what is known as a 'hierarchy of control'.

Source: SWA (2018d), Figure 1, p. 7.

FIGURE 1.4 Examples of potential falls in the domestic housing industry

Hierarchy of control

The hierarchy of control varies subtly from trade to trade, and state to state; however, under the National Falls COP there are five levels: the first level is the preferred option which, put simply, means 'do something else'; the last level is the least desirable option and requires something like the use of **personal protective equipment (PPE)**.

Level 1: Undertake the work on the ground or on solid construction.

Level 2: Undertake the work using a **passive fall prevention device**.

Level 3: Undertake the work using a **work positioning system**.

Level 4: Undertake the work using a **fall-arrest system**.

Level 5: Implement administrative controls.

- Level 1 effectively means bring the work to the worker, rather than the worker to the work. For example, some construction companies will construct a trussed roof at ground level, and then use a crane to lift the finished roof into position. It also includes, however, solidly built work platforms.
- Passive fall prevention, as level 2 suggests using, effectively means creating a safe **working platform** from which all of the required activities can be completed. Scaffolding is the most usual example; however, there are other systems such as mobile platforms or roof surround handrailing.
- If a fixed or temporary work **platform** cannot be provided, the next consideration is level 3 and a work positioning system. Basically, this is some form of restraint (complying with Australian Standards) that prevents you from going beyond a specific point – that is, you can't reach the edge to fall off. (See 'Work positioning systems' later in the chapter.)
- Next comes level 4, an **individual fall-arrest system (IFAS)**, catchment platform or net. This is more-or-less a last resort consideration, and is to be used only after determining that it is clearly impractical to implement any of the other levels.
- The final level, level 5, applies when a risk remains after consideration of the previous levels. In such cases, you must implement administrative controls (such as the use of observers and/or time limits on the activity). Level 5 control should only be resorted to when all other alternatives have been determined to be inappropriate, impractical or otherwise unachievable.

Administrative controls in this context means that – when no clear practical approach from levels 1 to 4 exists – documents stating who, when, where and how should be created. In this manner, we can clarify in everyone's mind that the chosen approach has been carefully considered as the safest way to go about a particular task. While such documentation should be in place regardless of the level of control used, its use here implies that it is the only control left – that is, make a plan and stick to it; improvise at your peril.

Ladders

Although considered part of Level 5, ladder use is addressed as a separate section (Part 8) in the National Falls COP. This is due to the number of serious and fatal injures occurring annually from ladder falls – even at very low heights.

The use of ladders is therefore always considered a last resort, rather than the logical and simple first choice that they are often (mistakenly) considered to be. In other words, we use ladders because there is no other way, not because it is the preferred way.

The type and load capacity of ladders to be used in any given context is clearly stressed in the National Falls COP, for example:

- minimum 120 kg industrial use load rating
- non-metal ladders for or near electrical work
- step and platform ladders not to be used for access to, or exit from, a work area.

Administrative controls should be developed when ladders are considered necessary. This ensures workers are approaching tasks safely and are not exposed to undue tiredness or strain.

KEY POINTS

- Ladders are your last resort – not your first thought.
- If you don't have to be up there, don't be!

Choosing a level of control: the concept of 'reasonably practicable'

Choosing a level of control requires careful consideration of the concept known as '**reasonably practicable**' (see SWA, 2022a, b). This means that, in making your judgement call, you would consider the following:

- the likelihood that someone might injure themselves or others in doing this task
- the severity of the injury that might be incurred in a fall or other possible incident
- the availability of the technology, systems, methods or equipment required for any particular level
- the frequency and/or duration of use of the control envisaged, and the longevity of that control. That is, will it need to be put in place once for a short time and then removed, be repeatedly installed and removed, be in place for an extended period of time, or remain in place after construction is completed? How long can the technology or equipment being used last in each circumstance?
- what is the cost? The harsh reality is that each control has a cost implication, but so do people's lives. It becomes a matter of 'weighing up' the various factors listed previously with the cost implications and the likely recovery of those costs. Cost most certainly doesn't come first, but neither can it come last; it is interwoven into your overall evaluation in determining whether something is 'reasonably practicable'.

Judgement-making is often a juggling act, rather than a linear or step-by-step approach. You effectively 'toss' each of these considerations back and forth alongside the level of control options available to you.

ON-SITE

EXAMPLE

You are required to install timber lining boards to the underside of a sloping ceiling in a small church (Figure 1.5). The boards are to be stained and oiled. Wall height to the ceiling is 3.6 m and to the ridge it is 5.4 m. For the most part, you will need to work at height while fixing material over your head.

FIGURE 1.5 Sloping ceiling in a small church

How your thinking might look on paper:

- I can't bring the ceiling to me, so I must go up to it to fix the boards (level 1 hierarchy of control).
- I'll check with the painters, but I'm sure they can stain and apply one coat of oil to the boards prior to fixing. This can be done on the ground.
- It's a one-off job, but the painters have to be able to get up there to do the final oil coat.
- I've got some trestle ladder frames and planks in the trailer, so I could use those, I suppose.
- Problem is, someone could easily fall (as we'll be looking up a lot of the time), and the severity of such a fall could be death or significant injury. I'd also have to move the **scaffold** frequently.
- What about a light, mobile scaffold...?
- I'd need to climb up and down from it to move it from place to place. And I need to be able to move easily along the full length of the board for fixing. It's also harder to sight along the length of the boards to check if they're straight. The painters may also find it difficult, as they would have to move the scaffold the length of the hall as they paint each board. I would have to hire the scaffold for longer, and the quality of the job could be compromised.
- Scissor lifts and other **elevating work platforms (EWPs)** are out of the question, as I'd need at least two (one for each end of the board) and I'd still have limited room for storing boards during fixing. Plus, we'd be constantly manoeuvring them as we fix. The job would be slowed, and I'd still have to keep them there for the painters to use. The hire cost would be high as well.
- A full scaffold deck running the length of the room would be better. Appropriately edge protected, this would lead to quicker work practices and a shorter hire period. It could also be used by the painting team. In addition, the end product would probably be of a higher quality.

In following this train of thought, you have covered areas of cost, frequency and duration of use, and availability, and you have moved through the hierarchy of control. At level 1 of the controls, you have done what you can do on the ground; you then moved up to level 2 and installed a safe work platform. In addition, you have considered worker satisfaction and, ultimately, the quality of the end product.

Identifying task requirements, work area and points of access

This is where you and your co-workers must determine from the plans and specifications just what it is that must be done. It fits in with the first stage of your hierarchy of control: Do we need to be 'up there' at all, or can the work be brought to the ground – for example, assembling a trussed roof on the ground and then using a crane to position it afterwards?

Assuming that you *do* have to go 'up there', what are the implications and the parameters? In discussing these, you should consider the following (see also Figure 1.6):

- What defines the work area?
- How will you access it?
- Is the existing structure capable of supporting the required fall protection?
- How will the materials be hoisted and positioned?
- How can you limit access to only those who need to be there?
- Do you need to put in public and vehicular barriers?
- What training and/or information is required for workers and supervisors?
- Is the work area unstable, fragile or brittle?

The last point noted here is perhaps the most important to consider when assessing the likelihood of a fall occurring.

Some old roofs where asbestos sheeting was used can be incredibly fragile and in most cases can not be accessed due to the risk of falling through. In these cases, it is often likely the roof will need to be replaced to resolve the access issue.

Another hazard that exists is when people paint roofs to 'spruce' them up. This hides the potential damage or weakness beneath and presents a hidden risk to people accessing the area.

Defining the work area	Where exactly do the workers need to be able to get to, and what will be the points of access and egress? Where are the areas of greatest risk (of falls, or of injury to those below)?
Consultation	Consult with the people who are involved in the work and site safety personnel to assist in making these decisions.
Providing access	How will workers access the work area? How will tools and materials be taken up and, if need be, where will they be stored?
Support of fall protection	On occasions, you may have to get an engineer's report, clarification from the site engineer or foreperson, or other certification regarding the capacity of the existing structure to support edge protection or to carry specific loads, or the appropriate load rating of an existing **anchorage point** (see 'Anchorage points, static lines and rails' later in the chapter).
Limiting access	How can you stop or limit unnecessary access? (**See Figure 1.7.**)
Vehicular barriers	Do you need barriers at ground level to stop vehicles or machinery (such as cranes) from hitting scaffolding or other elements of a work platform? Do you need signage and/or spotters, traffic wardens or the like?
Public barriers	Do you need a non-climbable fence or barrier to stop unauthorised access? Do you need to put the barrier in place so that people won't be hurt if tools or materials fall accidentally?
Information, instruction and training	Do the workers and/or other site personnel need training, special instructions or information regarding either the task, or accessing or working in the workspace?

FIGURE 1.6 Task requirements, work area and points of access

Source: iStock.com/Martti Salmela

FIGURE 1.7 Example of limiting access signage

Identifying appropriate fall prevention controls

Part of determining your level of control is considering all your options – basically, having a look at what needs to be done and where, and then doing some research to find out what equipment, methods or approaches are currently available to you. As always, those methods or approaches first considered should include ways of not having to be 'up there' at all.

In the next section, a basic overview is provided of the more well-known and commonly used systems available. These systems have been clustered in relation to their relevant hierarchy of control. The ceiling lining example offered previously assumes an experience base that, as a young tradesperson, you may not yet have. Even experienced tradespeople need to do their homework at times, as ideas, materials and equipment are constantly being improved.

ON-SITE

EXAMPLE

In the ceiling lining example offered previously, the issue of informing other workers who may need to enter, or work in, the church has not been considered. Nor has the issue of ensuring that all contractors have received training to work at heights. The timing of the work within the overall project schedule therefore needs careful consideration at the outset.

COMPLETE WORKSHEET 1

LEARNING TASK 1.2

1 **Circle 'True' or 'False'.**
Determining what is reasonably practicable is ultimately a question of cost.
True False

2 **Of the five levels within the hierarchy of control, the most preferred is:**
a Level 1
b Level 2
c Level 3
d Level 4
e Level 5

Maintenance and servicing

As with any equipment used in the construction industry it is important that all items are selected, inspected and maintained in accordance with the manufacturer's instructions and operation manual.

When working with equipment that has been damaged or is not fit for purpose the chances of a serious accident or fall from height are increased. PPE; for example, safety harnesses, must be inspected for any signs of damage or issues that may render the harness unsafe.

Even if a harness gets something like paint on it the manufacturer may state the harness is now unfit for use and must be replaced.

Always check the manufacturer's instructions in relation to using PPE that is designed to save your life in the event of a fall.

If you observe something that does not meet the manufacturer's specification with any fall from height equipment, including PPE, you must immediately notify your supervisor, tag the item out and remove the equipment from use so others don't use it.

Isolating the work area

Whenever people are working at height there is always potential to accidentally drop a tool or piece of material. Even with a roof railing or scaffolding in place, it is hard to stop items falling and this is a significant risk for people working on the ground.

To eliminate this hazard the space below the area at height should be isolated so other people do not access the area. This can be achieved by barricading the area using tape or orange/red mesh or other types of barricading that are available.

Signage should also be installed notifying people about the work occurring overhead and not to access the barricaded area.

The next sections provide information on the level of controls and how these can be implemented on a construction site to reduce the chances of a fall.

Consulting with relevant personnel

Remember the most important part of implementing any of these systems is ensuring they comply with the manufacturer's installation instructions and any relevant codes of practice.

Whether it is installing a hand rail around a roof or fitting a safety harness or installing a catch net below the work area, it is essential that you consult with relevant personal to confirm the fall protection requirements and equipment and that PPE is correctly installed and fitted before work commences.

Communication while working at height

Another important aspect to remember when working at height is to ensure good communication is maintained with all people involved in the work.

This does not only apply to the other people working at height but also applies to people on the ground, ensuring issues are reported and people are aware of their obligations while work is taking place at height.

Communication from the supervisor is also very important. They should monitor any changes in the weather that may have an adverse affect on the ground conditions or result in wet or windy conditions that could put the people working at height under increased risk of potentially slipping or falling from the roof area.

Adjusting and monitoring fall protection measures

Constructing or wearing a safety harness before work commences should always be seen as just the start of the fall prevention process.

Depending on the work flow you may need to adjust the controls that have been implemented to ensure they are fit for purpose in accordance with the manufacturer's specifications and operation manual.

For example, you may need to relocate or install another anchor point so you can access another section of roof. Or a scaffolding may need to be altered as the roof structure has changed through construction taking place and it is no longer useful or serving its designed purpose in its current position.

Inspection of worksites should be carried out on a routine basis to check how the construction progress may affect a fall prevention measure that has been put in place. For example, can the roof gutter be installed with a roof railing still in place and if it can't be, what is the work around?

Always ensure before any modification occurs for any fall protection control that you consult with relevant people onsite. For example, if you modify a scaffolding to suit your needs of installing roof sheets onto a roof, it may disadvantage or put at risk the bricklayers that are also using the scaffold.

Levels of control

Level 1 controls: solid construction

The National Falls COP defines 'solid construction' as:

- being structurally able to support workers, materials, and other associated live or dead loads
- having an even, non-slip surface and an appropriate gradient (generally not more than 7 degrees)
- having appropriate edge protection to all perimeters and penetrations
- having points of safe access and egress. (See **Figure 1.8**.)

Source: Advance Warehouse Structures

FIGURE 1.8 Example of solid construction

To satisfy each of these points, the following requirements must be adhered to in their entirety:

1. *Structural strength:* Both the working surface and the supporting structure must be designed with all conceivable loads (dead or live) allowed for. All proprietary branded props must be clearly marked with their **safe working load (SWL)** limits. All props must be tied and braced to prohibit movement.
2. *Work surface and gradient:* Must be non-slip and free of trip hazards. Generally, surfaces should be at a gradient of not greater than 7 degrees (1 m of rise to every 8 m of run). Gradients of 1 m of rise to 3 m of run (20 degrees) are allowable when cleats are used to increase grip (**Figure 1.9**).
3. *Edge and void protection:* Must be constructed to meet AS/NZS 4576 and AS/NZS 4994. That is:

FIGURE 1.9 Cleats on graded access ramp with gradient of greater than 7 degrees

   - it must be robust so as to withstand persons falling against it
   - handrails should be approximately 1 m above the work surface (between 0.9 m and 1.1 m)
   - it must be complete, with **midrails** or mesh infill, and **toe boards** (**Figure 1.10**)
   - it must have similar fall protection measures (such as gates) to all access points, ramps or stairs
   - voids, such as stairwells or partially completed floors (**Figure 1.11**), must have covers installed that cannot move and are capable of withstanding falling materials and/or persons.

FIGURE 1.10 Example of handrail with toe boards and midrails

4. *Access and egress:* Where using existing floor levels or permanent platforms, stairs, ramps or ladders, these must have been constructed to their relevant Australian Standard. All temporary access must be structurally sound, capable of withstanding all conceivable live and dead loads, and securely tied to the work platform.

FIGURE 1.11 Examples of voids

Note: Step ladders, trestle ladders, saw stools and the like *must not be used* for accessing or exiting a solid construction platform.

LEARNING TASK 1.3

1 **Circle 'True' or 'False'.**
Saw stools and step ladders may be used to access work platforms provided they are used sensibly.
True False

2 **Work surfaces should be of a gradient not steeper than:**
a 1:3
b 1:5
c 1:8
d 1:12

Level 2 controls: passive fall protection

Passive fall protection is a temporary system or structure that, upon being installed or brought into position, allows workers to complete their duties without having to further operate it for the duration of a specific task. This generally means scaffolding, EWPs, step platforms, trestle scaffolds and handrailing.

Chapters 3 (on EWPs) and 4 (on scaffolding) of this text cover these forms of passive fall protection in detail. For the purposes of this chapter, however, it is useful to offer here a brief overview of their types and forms.

Scaffolding

Rated as light duty (maximum load of 225 kg per bay), medium duty (450 kg per bay) and heavy duty (675 kg per bay), scaffolding is a system of interlocking components that provides a safe and stable working platform (Figure 1.13). Depending upon the duty rating, the working platforms provided will be from 450 mm (minimum) to 900 mm or greater in width.

Scaffolding is available in a variety of proprietary systems generally made from steel or aluminium, though bamboo is commonly used equally successfully in other nations.

ON-SITE

SOLID CONSTRUCTION AND ROOFING PLUMBERS

Plumbers working on roofs should be very careful when considering aspects of an existing roof or roof elements (such as box gutters and valleys – Figure 1.12) as 'solid construction'. Box gutters are prone to rusting out, and/or their supporting members may have rotted away. In such cases, using these roof elements as walkways can lead to an exaggerated sense of safety when, in fact, a highly dangerous condition exists.

The dangers of a fall aside, walking valleys – and particularly box gutters – can lead to the undetected failure of seams and joins due to the point load exerted by work boots and the like. Undetected, that is, until the next time it rains …

Source: Courtesy of Owen Smith

FIGURE 1.12 Box gutters and valleys

Source: Safe High-Ts Australia Pty. Ltd.

FIGURE 1.13 Scaffolding

All scaffolding must comply with the following Australian Standards (AS) set:

- AS/NZS 1576 Scaffolding (comprising six separate standards)
- AS/NZS 4576 Guidelines for scaffolding.

Unlike some other areas of construction safety (such as safe approach distances to power lines), there is national agreement as to who may assemble and/or install scaffolding. The *National Standard for Licensing Persons Performing High Risk Work* (Australian Safety and Compensation Council, 2006) sets out the following requirements:

- *where the scaffold's highest working platform is less than 4 m:* No licence is required; however, scaffold must be installed to the applicable Australian Standards by a **competent person** (a person trained and knowledgeable in the task and risks involved)
- *where a scaffold's highest working platform is greater than 4 m:* Depending upon the type of scaffold used, and its location or form, one of the following licences will be required:
  - Basic Scaffolding
  - Intermediate Scaffolding
  - Advanced Scaffolding.

All three scaffolding licences may only be obtained by persons over 18 years of age.

In addition to the above, scaffold must be inspected on a regular basis by a **licensed** scaffolder (generally at a maximum of 30-day intervals). (For more information, see 'Inspecting and maintaining the scaffold', in Chapter 4 of this text.)

Other issues surrounding the safe installation of, and work on, scaffold include:

- SWL, including live and dead loads
- access and egress (getting on and off)
- safe assembly and disassembly processes
- distance and/or protection from power lines (varying upon the voltage of the lines)
- unauthorised alterations to the scaffold
- appropriate edge and void protection
- stability and tying into existing building frameworks.

Scaffolding need not be static, however. **Mobile scaffolds** are also available that, in addition to complying with all of the above, must:

- remain level and plumb
- be kept clear of openings
- have lockable wheels
- not be accessed until all wheels are locked
- never be moved while someone is on the scaffold
- have all tools and materials appropriately restrained or removed prior to being moved.

Ultimately, it is the principal contractor of a work site who is responsible for the safe installation and maintenance of scaffolds. Likewise, the principal contractor must ensure that appropriate training, information and advice are provided to all those working on or around scaffold.

Note: Only a trained and/or licensed **scaffolder** *may add or remove components from a scaffold.* You must NEVER add or remove components yourself in an attempt to improve access or to store materials, or for any other reason. This can lead to a dangerous situation even if your intention is sound.

For further information on scaffolding, see Chapter 4.

Trestle scaffolds

A very basic form of scaffold is the use of trestle ladders (or other trestle-like frames) and planks to form a working platform not less than 450 mm wide (**Figure 1.14**). Generally suitable only for light-duty tasks, this form of scaffold has recently been brought into the scope of AS/NZS 1576. **Trestle scaffolds** must include handrails and midrails, and preferably outriggers, when the possible fall height is greater than 2 m.

Source: Quick Ally Scaffolding & Access Solutions (https://quickally.com.au/trestles-and-planks/safety-system)

FIGURE 1.14 Trestle scaffold with outriggers and handrail

Note: There are a number of issues with the use of trestle scaffolds that must be considered before deciding to use them. These include:

- Access and egress: How are you going to get up to the platform? (You cannot climb trestle frames, for example.)
- Materials and tools to be supplied to the platform: How much do they weigh, and how will you get them up there?
- How many workers will need to be on the platform? (Generally, a maximum of two.)
- What distance will the platform be from the work surface? (It must be narrow enough to prevent a person from falling through.)

For further information, see 'Trestle scaffold' in Chapter 4.

Step platforms

These are like an advanced step ladder: they are more stable, have a caged work platform at the top, and sometimes even a tool bench (see **Figure 1.15**). Made of aluminium, they are light and easily relocated and much safer to work from than a ladder. The main risk with their use is in becoming over-confident about your position and stepping backwards. This can lead to at

FIGURE 1.15 Step platforms

least a partial fall, even if you have the safety bar in place. (Some step platforms have a fixed bar that you must climb under to enter the cage.)

Step platforms are limited, however, in that they require good, level ground to operate from, which generally means working off a concrete slab or level work platform. All step platforms must comply with the AS series AS/NZS 1892 Portable ladders.

ON-SITE

SCAFFOLDING: A CAUTIONARY NOTE

While scaffolding, trestle scaffold and solid construction may be tied into an existing structure; however, great caution must be taken when doing so. The existing structure must be checked carefully for structural stability and integrity. In some instances, an engineer's report may be required. Chimneys, for example, particularly old ones, should never be used as a means of support or bracing, as they are notorious for failing under lateral loads. Likewise, brick and blockwork walls should not be used. Wherever possible, scaffolding and solid construction should be self-supporting and independently braced. Again, be wary of using chimneys and old block/brick-work for edge protection: such structures can fail just by leaning against their higher elements.

Using hessian or shade cloth as additional containment for falling materials should only be applied to scaffold specifically braced and restrained for such purposes. This is due to the increased wind loads associated with such materials acting as sails.

ROOFING PLUMBERS

Plumbers should take particular care around chimneys, old parapets and street-front façades, no matter what material they are made from. These are particularly unsafe as edge protection and should always be considered suspect.

Elevating work platforms

As stated earlier, Chapter 3 of this text covers EWPs in some depth. For the purposes of this chapter, a basic overview is all that is required.

EWPs are effectively an edge-protected, or 'caged', platform that can be raised, lowered or otherwise positioned so as to provide a safe work surface at heights. Scissor lifts (Figure 1.16), boom lifts (Figure 1.17) and vertical lifts all go under the heading of EWPs.

Source: Ahern Australia

FIGURE 1.16 Scissor lift

Source: Ahern Australia

FIGURE 1.17 Boom lift

(*Note:* Mast-climbing tower lifts are specifically excluded from this category.) AS 2550 Cranes, hoists and winches – Safe use – General requirements clusters EWP types as follows:

- trailer-mounted boom lifts (TL)
- self-propelled boom lifts (BL)
- vertical lifts (VL)
- scissor lifts (SL)
- truck-mounted boom lifts (TM).

Some EWPs are self-propelled (Figure 1.18), allowing them to be driven around a workspace. These may be battery driven, or powered by means of a small petrol or diesel motor. In addition, some self-propelled EWPs are suitable only for flat surfaces (such as concrete slabs), whereas others are adapted for use on rough terrain.

Source: Shutterstock.com/Hiper Com

FIGURE 1.18 Self-propelled elevating working platform

All EWPs on work sites must be designed, manufactured, maintained and operated with regard to the AS series AS 2550. A licence, issued by your relevant WHS/OHS authority, is required to operate any boom-type EWP capable of reaching 11 m or greater in height. Currently, vertical lifts (without a jib: a hinged or luffing member) and scissor lifts do not require a licence despite the vertical height they may be capable of reaching (National Falls COP).

Irrespective of the type of EWP or the height to which it may reach, *all* workers must be fully trained in the safe use and operation of the specific unit to be used before entering or operating that equipment. This training must be documented by the employer and signed by all parties. Training will include:

- selection and positioning of the EWP
- safe access and egress to and from the device (generally at ground level, unless AS 2550.10 is strictly adhered to in regards to its use as a means of delivering workers and equipment to an otherwise inaccessible work area)
- pre-operational and functional checks
- applicable PPE
- operation and maintenance of the equipment
- load and reach capacities
- post-operational checks
- emergency descent procedures
- distances from electrical hazards.

As part of the applicable PPE, the training must include the correct fitting and use of safety harnesses and any other fall-arrest or escape equipment.

Safety harnesses and EWPs

Although EWP platforms are fully edge protected, the work being undertaken frequently requires the user to lean to some extent out of the cage. In addition, the work platforms of cherry pickers, boom lifts and travel towers are generally small, while the height they are able to reach is significant. Further, these forms of EWPs can become unstable quickly with high wind gusts or sudden loss of hydraulic pressure. It is for these reasons that anyone working with such equipment must wear a safety harness or an IFAS; see Figure 1.19. This harness must be manufactured, maintained and anchored to meet the AS series AS/NZS 1891 Industrial fall-arrest systems and devices.

FIGURE 1.19 A safety harness or individual fall-arrest system must be worn when working on an EWP

Note: The IFAS must be connected to the EWP the moment you enter the platform. You *must remain connected* to the EWP at all times during the work and until you are about to exit the machine at ground level.

Further information on safety harnesses may be found in the discussion of IFAS under 'Level 4 controls: fall-arrest systems', later in this chapter.

Edge and perimeter protection

Edge protection was covered briefly under 'Level 1 controls' as part of a solid construction system. In this section, we will look at edge protection as a system of passive fall protection in its own right, as covered by the AS series AS/NZS 4994.1–4994.3 Temporary edge protection.

Note: AS/NZS 4994 is limited to roofs of 35 degrees or less. In addition, it presupposes that the standard will be applied to domestic construction and therefore to structures of not more than three storeys in height as per the *National Construction Code (NCC), Volume Two* (Australian Building Codes Board, 2019). For instances outside of this scope, the standard requires edge protection solutions to be specifically designed. Examples of these designs are offered in the latter part of this section.

Common examples of edge protection

Figure 1.20 shows a common example of proprietary bracketed edge protection for a domestic house. As with the edge protection discussed in the earlier section, there is a toprail and midrail, as well as a toe board. In this example, the pitch is less than 26 degrees. (A standard scaffolding example is offered in Figure 1.21.)

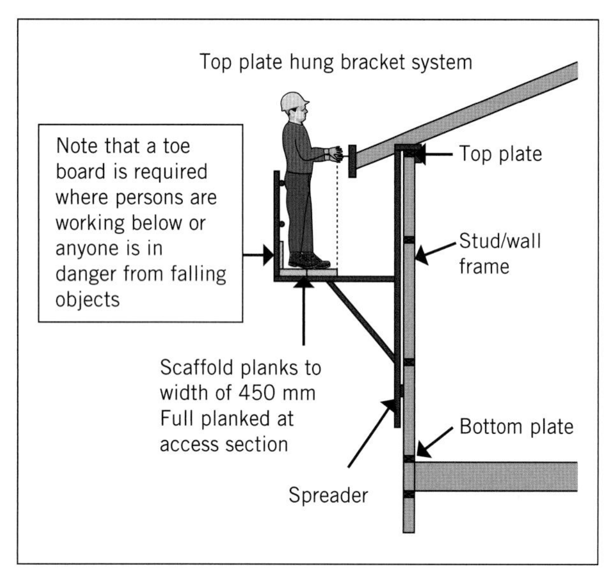

FIGURE 1.20 Proprietary bracketed edge protection for domestic house roof

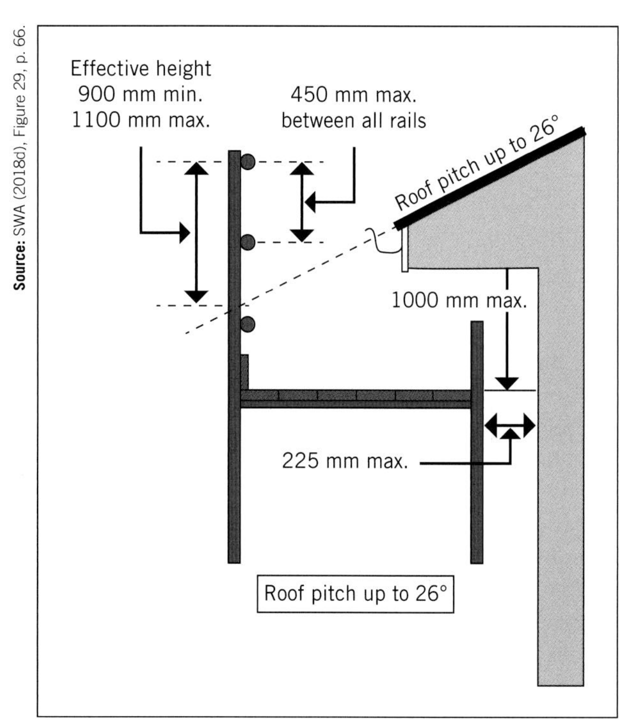

Source: SWA (2018d), Figure 29, p. 66.

FIGURE 1.21 Protection for roof pitches up to 26 degrees

Beyond 26 degrees, the standard (AS 4994) requires infill panels in addition to the toe board and midrail (Figure 1.22). This is due to the increased speed, and, hence, force, of someone falling against it. Note that the minimum width of the working platform is 450 mm, or the width of two planks.

The gap between the wall and the working platform must be narrow enough to prevent a person falling through.

ON-SITE

CONSIDERATIONS WHEN SELECTING EDGE PROTECTION

In deciding which edge protection is appropriate for a given situation, you should consider the following:

- *Roof surface:* A Colorbond® metal roof has the potential to be far more slippery (from early morning dew, or an unexpected shower of rain, for example) than a concrete tile roof in the same circumstances.
- *Roof pitch:* The pitch must be less than 35 degrees to be able to apply AS 4994, and less than 26 degrees to not require infill panels. Put simply, the higher the pitch, the faster the fall.
- *Roof slope length:* The longer the slope, the faster a person or object will be going before hitting the edge protection. More speed equals more force, and so the barrier must be that much stronger and better secured.

In addition to the roof type considerations, the following must also be considered:

- Can the structure to which the edge protection system is attached withstand the possible impact load of falling persons, materials or equipment?
- Does the system attach to rafters, trusses or wall framing?
- Are 'sacrificial brackets' (brackets that remain on the building after completion) required, or do brackets need to be fitted to trusses by the manufacturer prior to delivery?
- When can the edge protection go on, and when can it come off? (The system must be in place prior to any roof work commencing, and must not be removed until all roof work is completed.)
- How are you going to install the edge protection system? (This must be done in a safe manner that adheres to all the requirements already discussed in this chapter. This means you may need to use scaffold or an EWP to install it.)

In all cases the top- or handrail must be between 0.9 and 1.1 m in height above the working platform, or as measured 300 mm in from the finished roof edge (when installed close to the roof edge and without a work platform) (see Figure 1.22).

>>

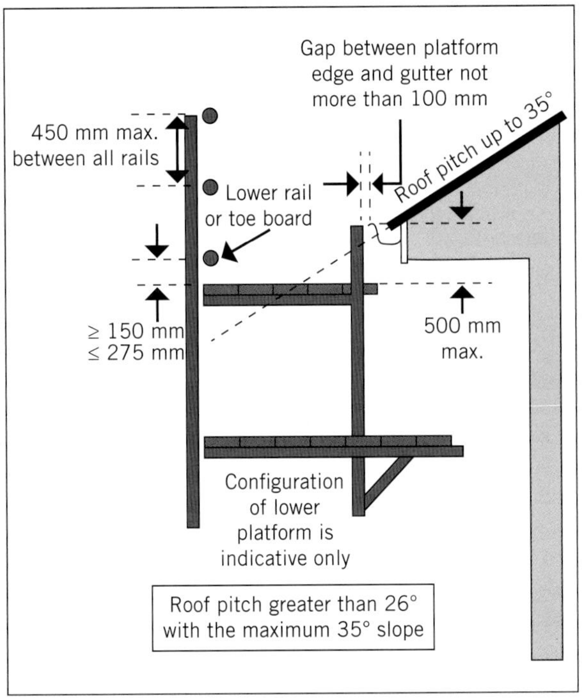

Source: SWA (2018d), Figure 29, p. 66.

Gap between platform edge and gutter not more than 100 mm

450 mm max. between all rails

Lower rail or toe board

Roof pitch up to 35°

≥ 150 mm
≤ 275 mm

500 mm max.

Configuration of lower platform is indicative only

Roof pitch greater than 26° with the maximum 35° slope

FIGURE 1.22 Protection for roof pitches up to 35 degrees

Edge protection for roof pitch greater than 35 degrees

Roofs with a pitch greater than 35 degrees are considered by the National Falls COP to be inappropriate to stand on and so require at least two of the following three options:

1. a work positioning system (see 'Work positioning systems' later in the chapter)
2. a roof ladder (see Figure 1.23)
3. an edge-protected scaffold platform (see Figures 1.21 and 1.22).

Source: Alamy Stock Photo/Alan Fern

FIGURE 1.23 Roof ladder

Option 3, supplemented by one or both of the other options, is the preferred system in most cases. Failure to achieve at least two of these options in tandem means that you will need to consider some form of EWP.

The National Falls COP suggests that all scaffolding used as fall protection should be erected for the height required as specified in the AS series AS/NZS 1576 Scaffolding.

In addition, scaffolding should be braced or tied to the building to ensure that it can withstand the lateral forces (force away from the building) that may be expected from a falling object hitting the platform and/or railing (see Figure 1.24). The higher the scaffold and the greater the pitch and length of the roof, the more critical this form of bracing or **tie** becomes. Again, for domestic construction it is assumed that a height equivalent to no more than three storeys or floors is required.

Note: While no licence is required to erect scaffold under the height of 4 m, it may only be erected by trained and experienced persons. Above 4 m, an appropriately licensed scaffolder must be used. Further information about scaffolding, its assembly and the relevant Australian Standards, may be found in Chapter 4.

Other perimeter or edge protection: guardrailing

Aside from the edge of a roof, there may be other areas that constitute a fall hazard to workers. These include skylights, existing fragile or loose roof materials, voids, window and door openings, stairwells, the edges of excavations, or simply the edge of a platform floor. Guardrailing is appropriate for these situations and, while there are proprietary systems available, it is common practice to construct this form of barrier from timber.

As with all other edge protection, **guardrails** must incorporate the following (Figure 1.25):

- top rails at 0.9 m to 1.1 m above the work surface
- midrails
- toe boards*
- installation according to the manufacturer's instructions, or constructed according to AS 1657 Fixed platforms, walkways, stairways and ladders.

AS 1657 provides specific material sizes for given spans and design features. However, as a guide, the timber sizes and spans set out in Table 1.2 may be used for general work.

The above is a guide only, however, and all timbers must be inspected by a suitably qualified and experienced person to ensure that they are suitable for guardrails. Knots, shakes and cracks, for example, can mean that the timber may fail under a sudden impact load from a falling person or object. Some timber defects (such as felling shakes) are particularly hard to

* *Note:* Where toe boards cannot be fitted, a 'no go' area must be created surrounding the work area to ensure that those below cannot be hit by falling objects. If the working platform has been created by a competent person, you must not attempt to 'jerry rig' (put together out of bits and pieces) something yourself.

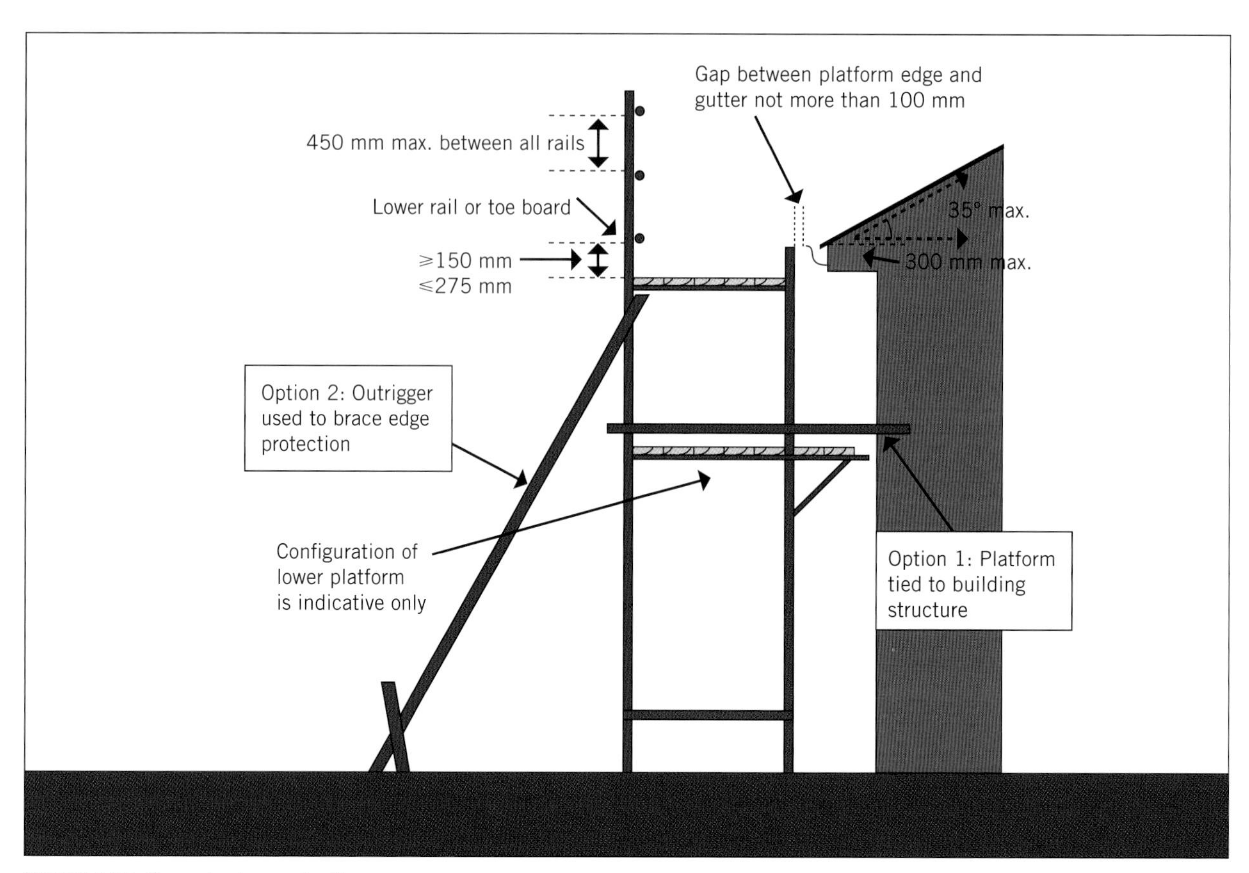

FIGURE 1.24 Braced-edge protection

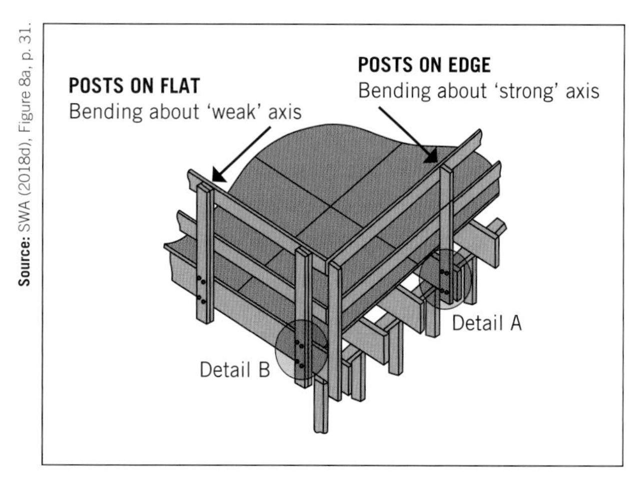

Source: SWA (2018d), Figure 8a, p. 31.

FIGURE 1.25 Example of timber edge protection showing bending around a weak axis (Detail B) and a strong axis (Detail A)

TABLE 1.2 Timber guardrails: sizes and spans

Rail size: depth × width (Nominal size in mm)	Span (m)	
	F8 Hardwood/MGP12 Pinus radiata	MGP10 Pinus radiata
100 × 38	2.7	N/A
100 × 50	3.5	N/A
2/90 × 35*	N/A	3.5

*Sections nailed together to form a 'T' or 'L', fixings at max. 300 mm centres

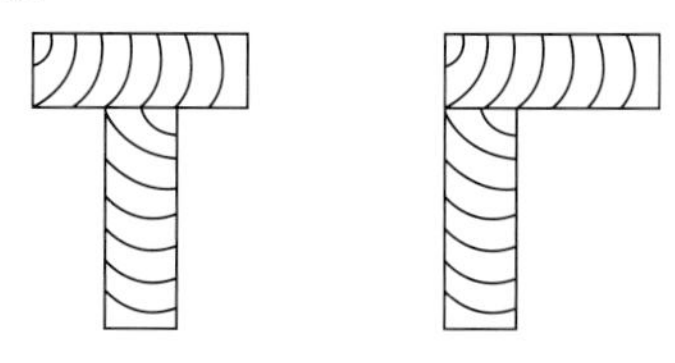

Source: Based on AS/NZS 4994.1: 2009.

spot and require an experienced eye to pick them up (Figure 1.26).

Timber posts should be fixed so that any impact load forces act upon the longest sectional size (their strongest axis: see Figures 1.25 and 1.27).

Figure 1.28 shows how this system might be applied to a stairwell opening on an upper floor level.

Note regarding trenches and guardrails

There is now a model code of practice covering trenches and excavation in general. The *Excavation*

FIGURE 1.26 Example of felling shake

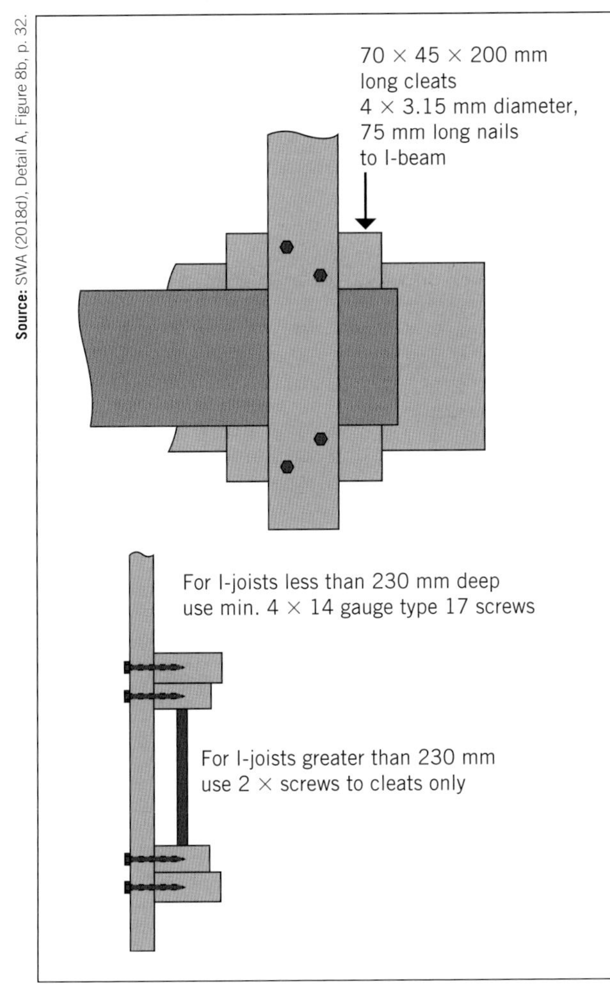

FIGURE 1.27 Fixing timber posts

Work: Code of Practice (SWA, 2018b) has been adopted by all states and territories in its main terms (and in its entirety by most states). This code requires that any trench or excavation into which a person may fall 1.5 m or more must be controlled by edge protection (though experience suggests that you should protect

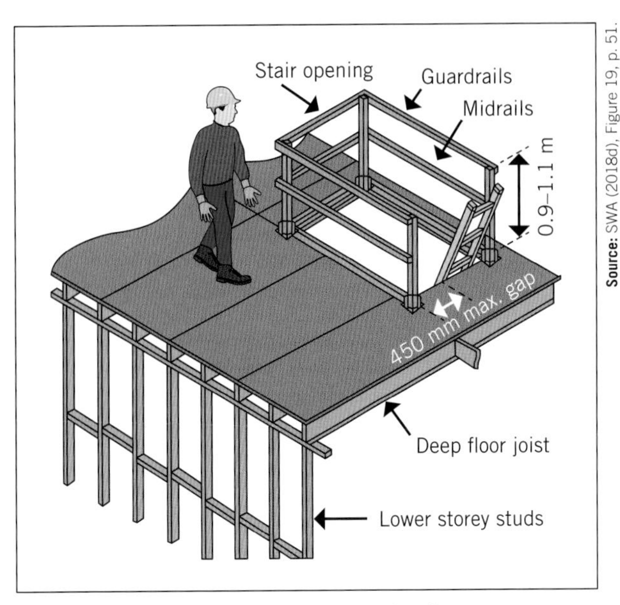

FIGURE 1.28 Edge protection around stairwell

all trenches, and guardrail anything greater than 1 m of fall).

Other considerations for trenches include:

- How may access and egress be achieved safely?
- Should intermediate platforms be used for deep trenches?
- Can the trench be **backfilled** progressively to reduce the length of open trench?
- Can the trench/work area be isolated by way of perimeter fencing, **barricades** or the like?
- Can trench covers be created, limiting the effective length of open trench?*

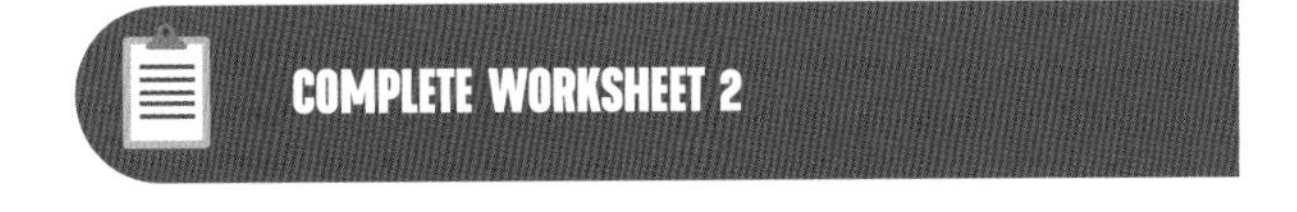

LEARNING TASK 1.4

1 **Circle 'True' or 'False'.**
The minimum width of any working platform is always 450 mm (two planks).
True False

2 **You must hold a scaffolding licence if distance from the ground to the working platform is equal to or greater than:**
a 2 m
b 3 m
c 4 m
d 5 m

* *Note:* Trench covers must be structurally sound to span the width of the trench with the possible impact loads taken into account. They must also be fixed into position.

ON-SITE

INSTALLING EDGE PROTECTION FOR TRENCHES

Unfortunately, trenches also tend to be rather difficult for positioning a guardrail close to the edge. You cannot drive posts into the ground, for example, as this could collapse the side of the trench. The approved practice is to extend trench shoring above the trench top edge by the required guardrail height (generally, 1.0 to 1.1 m: see Figure 1.29). Some proprietary systems allow for fixing to the top edge of the shoring, which effectively does the same thing (i.e. extend the shoring).

FIGURE 1.29 Guardrail to trench

Short-term barriers, such as post (capped star pickets, for example) and webbing, must be placed at least 2 m from the trench edge. This is because barriers of this type cannot withstand someone, or something, falling against them: they act as a visual warning and are not a guardrail.

Level 3 controls: work positioning and travel restraint systems

Though sometimes the terms *work positioning* and *travel restraint* are used to imply the same system, each is in fact designed for very different work situations. Despite this, each system is based upon a similar array of components.

Work positioning systems

A work positioning system (Figure 1.30) allows the worker to be held, suspended in position, by way of appropriately designed equipment. Generally, this implies the use of a full body harness and an array of (i.e. more than one) connecting devices such as pole straps, an

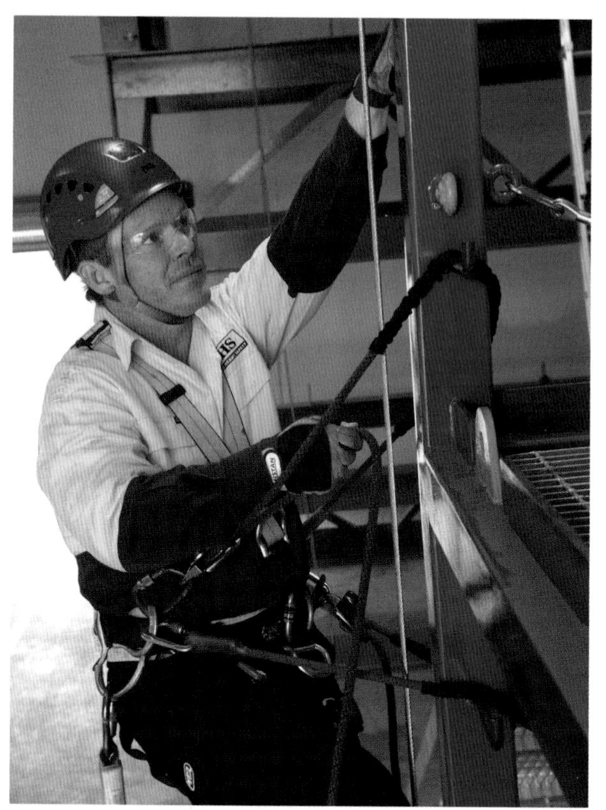

Source: Total Height Safety Pty Ltd

FIGURE 1.30 Work positioning system

energy-absorbing **lanyard** (used as an emergency fall restraint), an **industrial rope access system** or rail grab system, and/or a system of ropes and karabiners.

The purpose of a work positioning system is to locate a worker safely such that they can access the job and, most importantly, work with both hands free. The worker, having reached the work location by means of a ladder, EWP, rope or abseiling technique, can then be secured in a stable position. This security is based on a series of safety links to the structure so that, should one fail, another is there to limit any associated fall.

Travel restraint systems

Travel restraint systems, on the other hand, are a means of limiting the position or travel of a worker to a known safe area. In its crude historical beginnings, this was not much more than a rope that stopped the worker from getting too close to the edge of a work area. Contemporary systems now use a full harness (Figure 1.31) manufactured, maintained and anchored to meet the AS series AS/NZS 1891 Industrial fall-arrest systems and devices. The 'rope' has been replaced by a lanyard approved to the same standard, the length of which is determined by the distance to the nearest fall risk (Figures 1.32 and 1.33).

The simple premise behind a work positioning and a travel restraint system is that the person working at height is always connected to the roof. Depending on the layout this may mean that the person has two lanyards attached to a harness. The person can then attach to the closest anchor point, traverses to the next

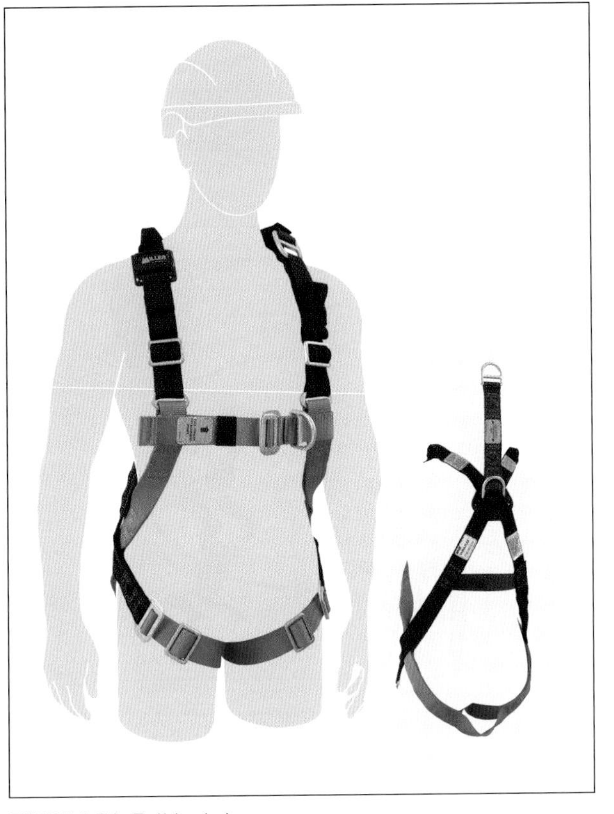

FIGURE 1.31 Full body harness

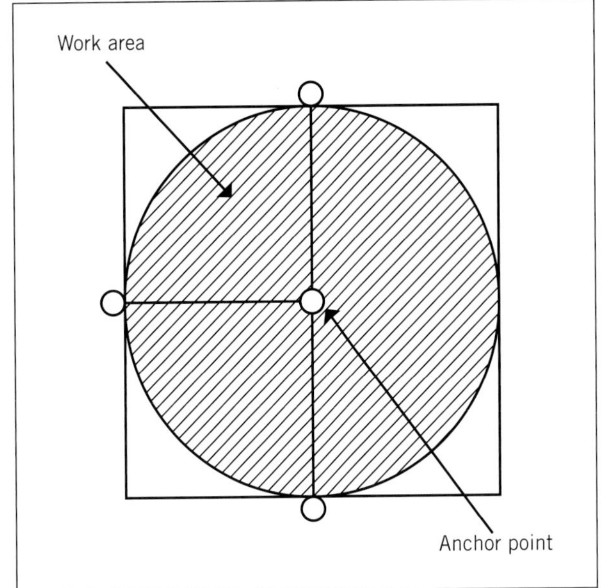

FIGURE 1.32 Limit of work area defined by travel restraint system

anchor point and secure the second lanyard to this, then returns to the first to disconnect. This method ensures the person is always connected to the roof.

Mistakes in installation and use are very easily made with travel restraints, which can lead to what is known as the pendulum effect (see Figures 1.34 and 1.35). For this reason, they should only be used under strict training and supervision guidelines.

Retractable lanyards and inertia reels (similar to a seat belt roll in a car) are prohibited with travel restraints because they may allow a worker to get beyond the safe work zone.

KEY POINTS

- Travel restraint systems are a level 3 control. This means you should only use such a system after determining that it is not possible to do the work from the ground, from solid construction, or by the installation of edge protection.
- Anchoring points designed and installed for travel restraints prior to recent changes in the standards generally cannot withstand the impact loads generated from a falling person. The restraint must therefore stop a person from reaching any point of the work area from which they might fall (see Figures 1.33, 1.34 and 1.35).

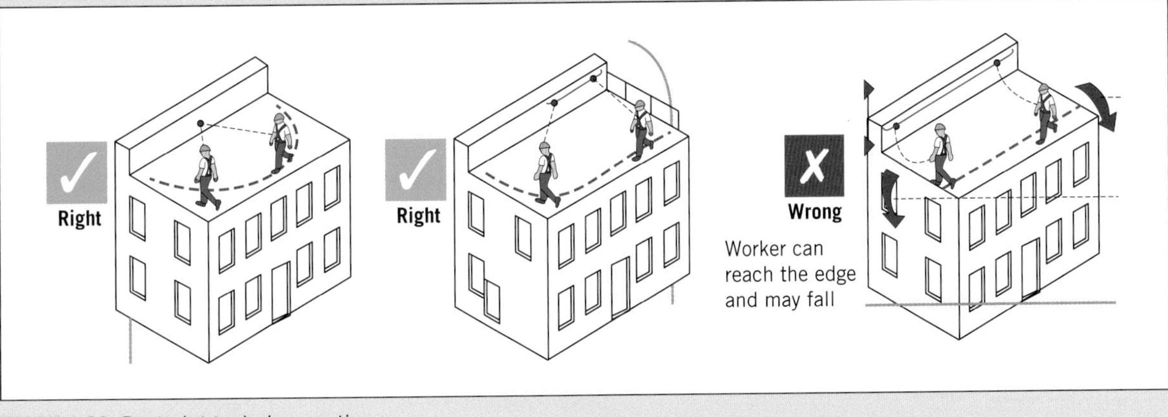

FIGURE 1.33 Restraint technique options

>>

- A travel restraint is not a fall-arrest device! The lanyard is of a fixed length and must not include an **energy absorber**. This is because the energy absorber is capable of effectively extending the lanyard beyond the safe work area. A genuine restraint lanyard of the correct length must be used.

Source: SWA (2018c), Figure 23, p. 51.

FIGURE 1.34 Pendulum effect: 1

Source: SWA (2018c), Figure 24, p. 52.

FIGURE 1.35 Pendulum effect: 2

LEARNING TASK 1.5

1 **Circle 'True' or 'False'.**
Travel restraint systems limit the position or travel of a worker to a known safe area, whereas a work positioning system locates a worker safely such that they can access the job and work with both hands free.
True False

2 **Circle 'True' or 'False'.**
With travel restraint systems, the lanyard is of a fixed length and must not include an energy absorber.
True False

Level 4 controls: fall-arrest systems

Unlike the previous levels, level 4 controls assume that, due to the nature of the work, a fall has become highly possible. Such being the case, you would only go to this level when all previous options have been eliminated as impractical – that is, the work cannot be done on the ground, a solid platform cannot be constructed, and scaffolding or EWPs cannot be used.

Because the possibility of a fall is so great, some form of 'fall-arrest' system is required. By this means the distance of a fall is reduced and the worker is prevented from falling all the way to the ground. Industrially, there are three ways by which we might achieve this:

1 catch platforms
2 safety nets
3 IFAS or safety harnesses.

Catch platforms

The term **catch platform** has been applied, and is still applied, to a variety of temporary structures around a building site. Generally, however, a catch platform falls into one of two types:

1 a temporary structure designed to protect the general public and/or workers below the work area from falling tools, debris or materials. It may also act as a raised walkway around the work area. A related term for this form of structure is **gantry**
2 a temporary structure immediately below a work area from which a fall is highly possible given the nature of the work undertaken. The structure is designed to withstand the impact load of a falling person, persons and/or materials.

The first description has been covered under level 2 controls (see 'Edge and perimeter protection'), and so it is the second definition that is of interest. A catch platform is effectively a raised floor underneath the work area (Figure 1.36). Catch platforms need to be constructed by competent tradespeople or scaffolders (depending upon the materials used) to meet the AS series AS/NZS 1891.

In brief, this means that the area must:

- be fully decked
- extend at least 2 m past the work area from which someone could fall. (This may be reduced if there is a guardrail fitted to the catch platform. The railing must extend a minimum of 1 m above the work area)
- be located such that it is no more than 1 m below the work area (preferably less).

Source: Images by GCS

FIGURE 1.36 Catch platform

Catch platforms may be constructed out of anything, so long as they comply with the standard (AS/NZ 1891) and, in so doing, can withstand all potential impact loads. Since impact load increases with the distance of a fall, the closer you can position the platform to the underside of the work area, the better your system. Scaffolding, trestle scaffold and braced acrow props are all logical components in such a system; however, timber may also be used.

Note: In New South Wales, catch platforms must only be constructed by those holding a rigger's and/or scaffolder's licence. The level of licence required will be determined by the components required to be used and the location of the scaffold (i.e. suspended).

Safety nets

Safety nets could be thought of as a flexible catch platform. Unlike the safety nets you may have seen at a circus (where the trapeze artist may fall several metres before landing safely in the net), workplace safety nets are positioned very closely beneath the area of work (Figure 1.37). This is because (unlike the circus performer) the average worker is not trained in how to fall safely from a height and 'tumble' into a net. More likely, the worker has missed their footing or has been knocked, and so the fall is sudden and unexpected. In addition, they (it may be you) may well be carrying tools or equipment, which will compound the risks associated with the fall.

Unlike catch platforms (or the gantry form), safety nets should not be considered safe to work under. Tools and materials could fall through them, and a significant impact load could stretch the netting such that a worker below may be struck.

Source: Alamy Stock Photo/lowefoto

FIGURE 1.37 Safety netting

Given the amount a net may stretch, the positioning of safety netting is critical. This is to ensure that sufficient room beneath the net is maintained. 'Sufficient room' is determined by the greatest load the net is rated for not impacting the floor or any objects between the floor and the net. Generally, this information is provided by the manufacturer.

Due to the critical positioning requirements, the complex 'mechanics' of safety netting, and the various connections and methods required to tie them off, installation must be carried out as per the manufacturer's instructions and by competent, trained persons: this generally means a rigger or scaffolder, unless the manufacturer has provided specific, certified training to other personnel.

Key considerations concerning the positioning, use of and work around safety nets are as follows:

- Safety nets should be located such that anyone or anything falling into the net will not strike objects beneath it (i.e. top plates of walls, lower floor surfaces, piping extending from floors or walls, and so on).
- Safety nets must not be deliberately jumped into, walked across or used as a place to store materials.
- Anything that falls into a net must be removed as soon as practicable. Failure to do so could overload the net, or injure someone falling into the net.
- You must have a recovery plan in place for the removal of persons or objects from the netting.
- Connection points must be approved for the load that the net is likely to impart to it (should an object fall into the net).
- The structure to which the net connection points are attached must also be rated for the probable loads. This means scaffolding may require additional bracing or ties.
- Welding, grinding or oxy-acetylene cutting must not be conducted above or near a net. (Safety netting is generally nylon and may melt or be otherwise degraded by sparks or flame.)

- Nets should not be used in situations where chemicals or high heat are likely to impact upon them.
- Safety nets must be inspected regularly, before, during and after installation.

Individual fall-arrest systems

Individual fall-arrest systems (IFAS) have become more common on building sites, but offer a mistaken sense of security. In addition, they are cumbersome to use and in many instances can impair a worker's capacity to work or move safely through a structure at heights, rather than improve it.

In recognising these and other factors (such as constant supervision, inadequate anchorage points, and the clear fall height required for them to fully deploy – over 6.5 m), the National Falls COP suggests that the IFAS has limited application in the domestic housing sector.

That stated, you will come across occasions when you will need to wear a harness if there is no other fall protection available – for example, when traversing a large sloping roof area that has a series of dedicated roof anchors so as to undertake maintenance or repairs. IFAS are also an important piece of equipment for plumbers sheeting out roofs. Further, timber framing is now acceptable to three-storey construction. Coupled with small blocks, yet clients still wanting large homes, potential 6 to 7 m falls are no longer uncommon on the domestic scene.

It is therefore important that you make yourself familiar with the following:

- the name and use of the components of a typical IFAS
- how to size and fit an IFAS harness appropriately
- the importance of correct and timely recovery procedures
- how to calculate the total required deployment distance
- the correct selection and application of lanyards and inertia reels
- risks and hazards associated with IFAS use
- recovery (rescue) plans and appropriate techniques.

Components of an IFAS

The components of a typical IFAS are shown in Figure 1.38.

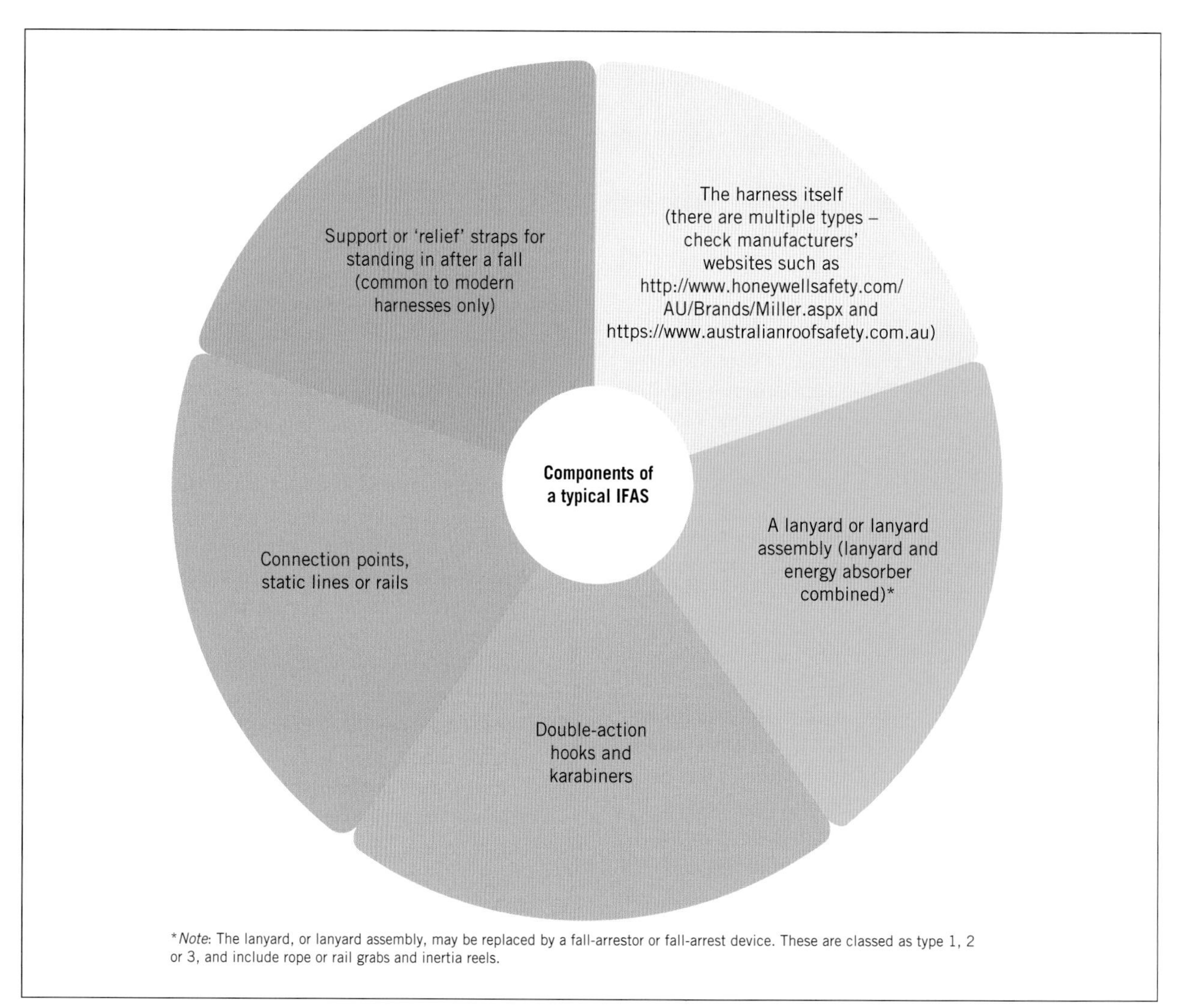

FIGURE 1.38 Components of an IFAS

Harness

This is the core of your safety system and must be chosen and fitted with care. The choice is wide; however, in the main, a harness comprises a system of webbing straps that go around the body, including between the legs and across the chest. Generally, they will have three attachment points (two in the front and one at the back) at which all the webbing clusters (Figure 1.39).

Ordinarily, the lanyard is attached to the 'D' ring of the rear cluster point. (Some harnesses will have a short piece of webbing already attached and ready to receive the lanyard.) This is because, in a fall, the lanyard is then less likely to strike your face or head. The front points are generally used in recovery operations or as part of a work positioning system (see, Figure 1.45).

All harnesses in Australia must be tagged (Figure 1.40) and show the following information:

- manufacturer's name
- date of manufacture
- withdrawal from service date
- material it is made from
- rated drop distance(s) (May vary for rear, front or side attachment points.)

Before use, you must always check the following:

- webbing:
  - labels are present, and are legible and include expiry date
  - cuts, frays, abrasions or damaged stitching
  - heat, chemical, paint or glue damage
  - rotting or mildew
  - failed or worn stitching
- fittings (buckles, 'D' rings, karabiners, hooks or latches):
  - distortion
  - cracks or breaks
  - wear of moving parts or clips.

FIGURE 1.39 Harness and 'D' ring

HOW TO

HARNESS FITTING

Figures 1.40 to 1.45 outline the correct procedure for fitting an IFAS. Once completed, it should be a firm fit without being constrictive.

1. Check the harness for any damage and the harness label for adherence to AS/NZS 1891. Make sure the harness and lanyard assembly have not exceeded their expiry date.
2. Hold the harness by the 'D' ring, as shown. Shake the harness and ensure that all the straps fall into place. Make sure the leg and chest strap buckles are unbuckled. (See Figure 1.41.)
3. Having identified the top rear 'D' ring, put the harness on much as you would a high-visibility vest. Once the shoulder straps are in position, ensure that the rear 'D' ring is central to your upper back. (See Figure 1.42.)
4. Adjust and fasten the chest strap. (See Figure 1.43.)
5. Pull the leg straps between your legs and link to the side buckles, as shown. Be sure not to let the straps get crossed or mixed up. The fit should be firm but not restrict movement. (See Figure 1.44.)
6. Run your hands flat over the webbing to ensure that no twisting or bunching has occurred. (See Figure 1.45.)

Should the harness not feel right, or if there is evidence of incorrect fitting, remove it completely and start again.

Note: You must always take ultimate responsibility for ensuring your harness is properly fitted. Whenever possible, however, have a competent person check your harness (particularly the back webbing) for twists and correct connections.

>>

>>

FIGURE 1.40 Check harness labels: step 1

FIGURE 1.41 Harness fitting: step 2

FIGURE 1.42 Harness fitting: step 3

FIGURE 1.43 Harness fitting: step 4

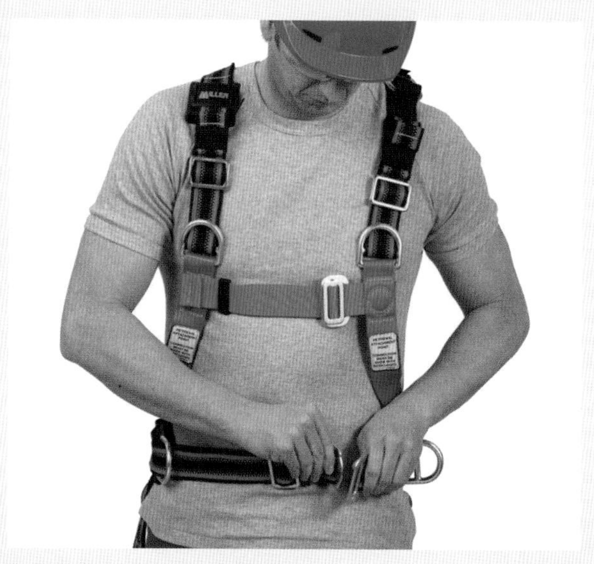

FIGURE 1.44 Harness fitting: step 5

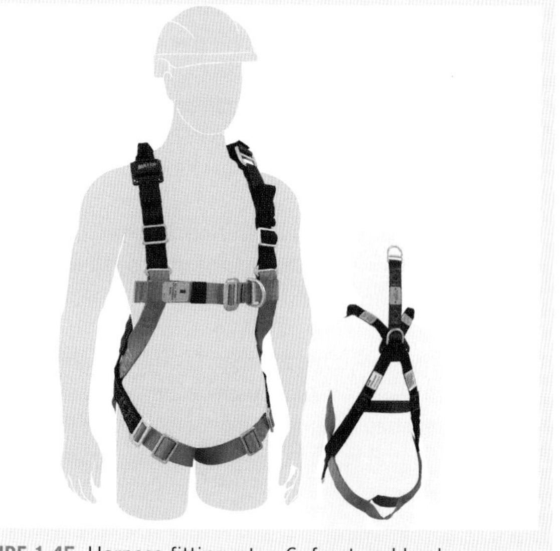

FIGURE 1.45 Harness fitting: step 6, front and back

Lanyards and lanyard assemblies

Lanyards may be made of rope (Figure 1.46) or webbing (Figure 1.47), and may be single or double (Figure 1.48) in form. Lanyard assemblies differ only in that they have an integrated shock or energy absorber (also called a 'deceleration device'). In addition, there are adjustable-length lanyards (Figure 1.49): these lanyards are fitted with a 'rope grab' device (see 'Fall-arrestors and devices' later in the chapter) by which the effective length of the lanyard may be reduced.

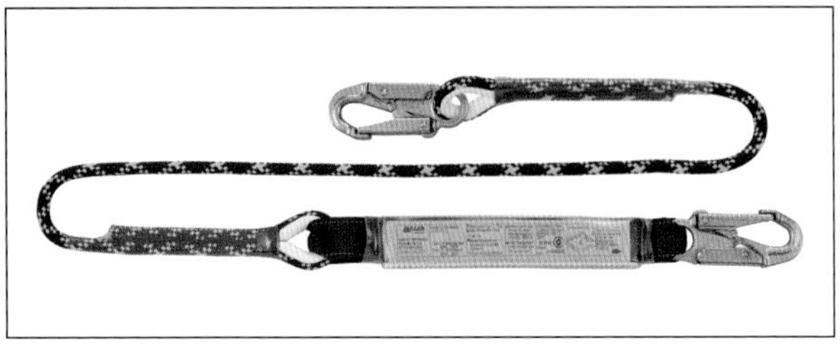

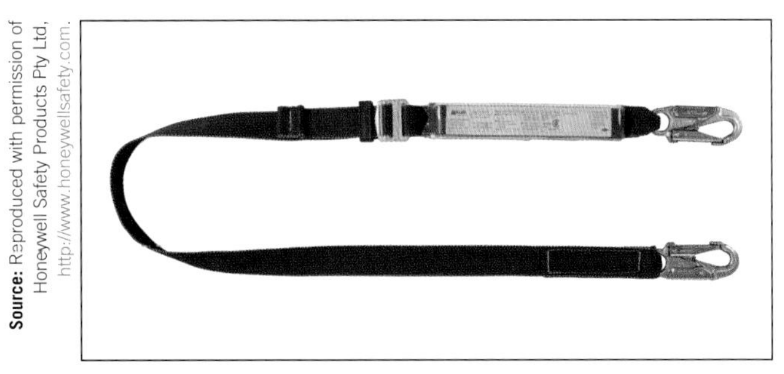
Source: Reproduced with permission of Honeywell Safety Products Pty Ltd, http://www.honeywellsafety.com.

FIGURE 1.46 Rope lanyard

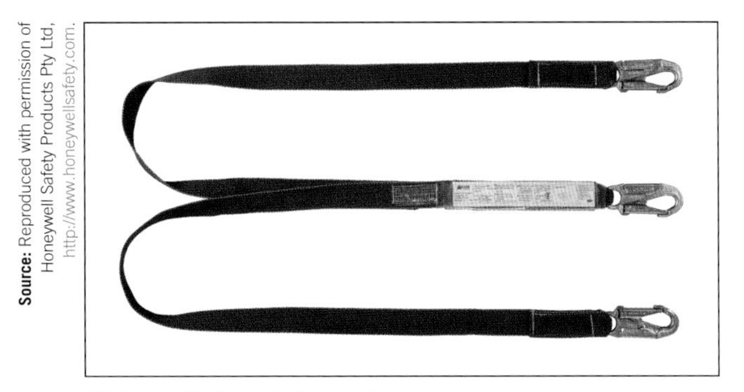
Source: Reproduced with permission of Honeywell Safety Products Pty Ltd, http://www.honeywellsafety.com.

FIGURE 1.47 Webbing lanyard

Source: Reproduced with permission of Honeywell Safety Products Pty Ltd, http://www.honeywellsafety.com.

FIGURE 1.48 Double lanyard

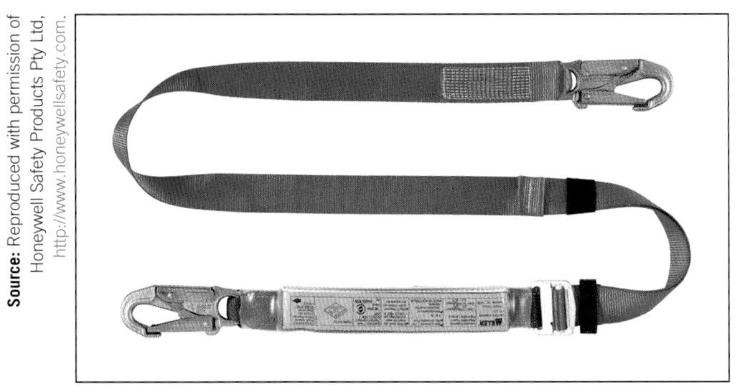
Source: Reproduced with permission of Honeywell Safety Products Pty Ltd, http://www.honeywellsafety.com.

FIGURE 1.49 Adjustable lanyard

It is important to be aware of the length of the lanyard, including the expanded length of the energy absorber (should it be fitted). As shown in Figure 1.50, it is from this information that the total clear activation distance is calculated. All lanyards must comply with AS/NZS 1891.1.

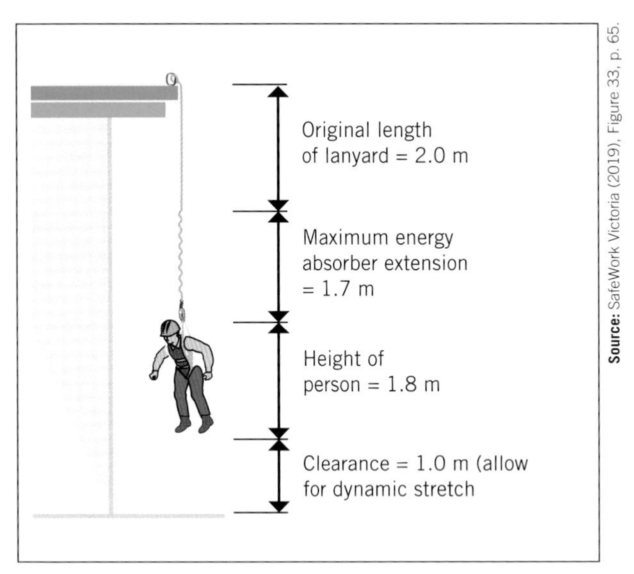

Source: SafeWork Victoria (2019), Figure 33, p. 65.

FIGURE 1.50 Total fall clearance required = 6.5 m

Fall or clear activation clearance

Generally, a lanyard is limited to 2.0 m in length, and the energy absorber expands out another 1.7 m. However, as shown in Figure 1.50, to this must be added the height of the worker (nominally 1.8 m) and at least 1.0 m more for the dynamic stretching of the webbing. That is, at least 6.5 m of clear fall distance is required to prevent someone from hitting the ground.

HOW TO

EXAMPLE

$$\begin{aligned}\text{Fall clearance} &= \text{lanyard length} + \text{energy absorber length} \\ &\quad + \text{worker height} + \text{stretch clearance} \\ &= 2.0 + 1.7 + 1.8(\text{nominal})^{*} + 1.0 \\ &= 6.5\text{ m}\end{aligned}$$

*In practice, this distance is actually shorter, based upon the location of the connecting 'D' ring at the back of the harness (at least 200 mm lower, making it 1.6 m). However, we tend to err on the side of safety and leave it at 1.8 m.

It is this extended distance that limits the use of IFAS in domestic construction because in single-storey domestic housing this clear fall distance will never be obtained. Failure to understand this requirement can lead to a false sense of security in the mind of the worker, which in turn can lead to injury or death.

Limited free fall
If a person falls from height and is caught by the harness, it will likely save their life by stopping them from hitting the ground. However, depending on the distance they fall before being caught by the IFAS, the impact the body receives will still likely cause injury. When setting up anchor points and lanyard lengths this should be considered. The amount of 'free fall' from the area at height to the point where they are caught by the IFAS should be limited as much as possible.

Fall-arrestors and devices
A fall-arrestor is a form of self-locking device that stops (arrests) a fall. These devices include rope and rail 'grabs' and inertia reels.

Type 1 arrestors. Rope and rail grabs (**Figures 1.51**, **1.52**, **1.53** and **1.54**) form this category. They are connectors designed so that they can easily be moved along a line or rail and yet, when a load is put on them suddenly (such as through *a person falling*), *they lock into position ('grab' the line) and so arrest any further movement.* Connection to a type 1 arrestor is by means of a short lanyard (length depending upon the application) or direct karabiner link to the activating lever. Rope and rail grabs are used in a number of settings, such as ladder fall-arrest systems (**Figures 1.51** and **1.53**) or horizontal guide rail systems (**Figure 1.52**).

Type 2 and 3 arrestors. These devices are **inertia reel** systems that, like a contemporary car seat belt, allow you to pull out slowly as much webbing as you need

Source: Reproduced with permission of Honeywell Safety Products Pty Ltd, http://www.honeywellsafety.com.

FIGURE 1.52 Horizontal guide rail

Source: Reproduced with permission of Honeywell Safety Products Pty Ltd, http://www.honeywellsafety.com.

FIGURE 1.51 Rail grab

Source: Reproduced with permission of Honeywell Safety Products Pty Ltd, http://www.honeywellsafety.com.

FIGURE 1.53 Ladder line grab

Source: Reproduced with permission of Honeywell Safety Products Pty Ltd, http://www.honeywellsafety.com.

FIGURE 1.54 Rope grab

(or it contains) and then automatically wind back any that you don't need. Provided that your movements are smooth and steady, the reel doesn't lock. Given a sharp tug, however, the reel locks up. It is by this means that a sudden fall is prevented or arrested before going very far. The advantage of this system is that you are not tripping over unnecessary lanyard length, yet still retain a great deal of movement. Standard inertia reels (Figure 1.55) are classed as type 2 fall-arrestors, while those with an inbuilt retrieval winch (electrically or hand driven) are classed as type 3 (Figure 1.56).

Again, it is important to understand the total fall clearance that is required for these units to be used safely. This can only be done by reviewing the

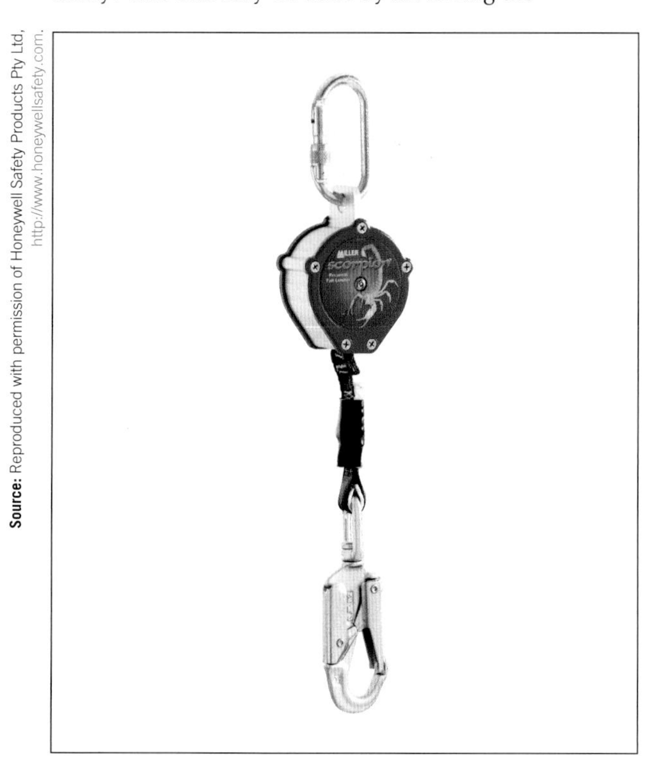
Source: Reproduced with permission of Honeywell Safety Products Pty Ltd, http://www.honeywellsafety.com.

FIGURE 1.55 Inertia reel

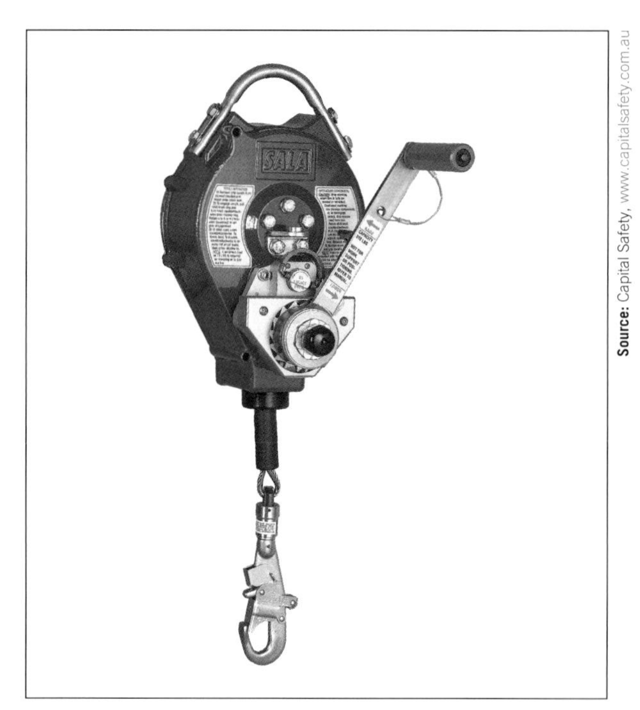

Source: Capital Safety, www.capitalsafety.com.au

FIGURE 1.56 Inertia reel with retrieval winch

manufacturer's instructions specific to that particular unit. Some arrestors have short travel energy absorbers built in as part of the webbing, for example. In such cases, while they may state that they stop a fall in mere millimetres, the unravelling of the shock absorber may add another metre or more to the clearance required. The inertia reel should always be located at the static connection point and not on the harness.

Hooks and karabiners

Karabiners (Figure 1.57) are a form of lightweight, but high-tensile (strong in tension), metal connector used in a variety of safety, climbing and rescue situations. In the IFAS, the karabiner is generally the connection point between your harness, and lanyard or other fall-arrest device, and the lanyard and anchorage point. AS/NZS 1891.4 states that karabiners must require two separate actions to open. Generally, this means that some form of

Source: Reproduced with permission of Honeywell Safety Products Pty Ltd, http://www.honeywellsafety.com.

FIGURE 1.57 Karabiner

screw or twist lock sleeve must be undone before the gate may be opened. Karabiners that do not have this system can fail should the gate open through inertia actions during a fall. (That is, the gate can 'swing' open, weakening the karabiner.)

Note: When locking this form of karabiner, do the screw gate up tight, then undo it a quarter of a turn. This is critical because, under a 'shock load' (i.e. falling), the screw can seize and you (or a recovery team assisting you) may not be able to undo it.

Hooks, most of which are a bit like a large karabiner by another name, must also have a double action to open. Generally, this is by a lever on the back of the hook, which is pressed by the base of the fingers or the palm of the hand (Figure 1.58). It is by means of this large 'karabiner' that you would generally connect to rails, static lines or anchorage points on the building.

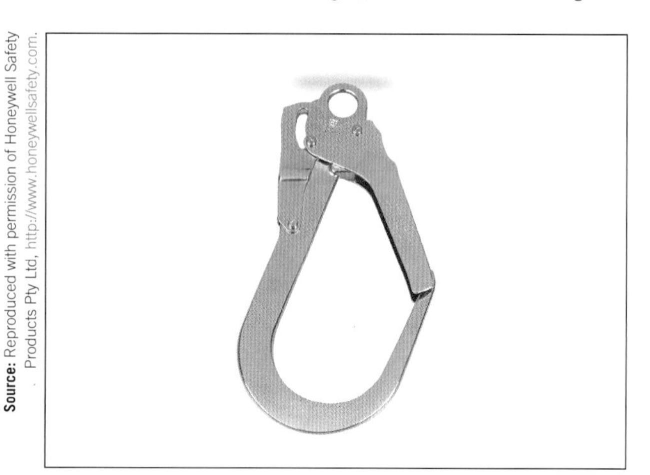

Source: Reproduced with permission of Honeywell Safety Products Pty Ltd, http://www.honeywellsafety.com.

FIGURE 1.58 Hook

Source: SAYFA GROUP, www.sayfa.com.au

FIGURE 1.59 Roof anchorage point

Source: Reproduced with permission of Honeywell Safety Products Pty Ltd, http://www.honeywellsafety.com.

FIGURE 1.60 Static line

Anchorage points, static lines and rails

While the body harness is the core of the IFAS, the whole system ultimately comes back to that to which you are connected. There are basically two categories of connection points: anchors (Figure 1.59), and static lines or rails (Figure 1.60).

Anchor points must be installed in a location that allows for easy access and complies with the related code of practice. To achieve this the reasons for accessing the area at height should be analysed. For example, air conditioners that are installed on a roof may have a anchor point installed in close proximity to the system to allow the service person to connect a lanyard while carrying out servicing work.

Connection or anchorage points (also known as anchor points) must be designed to the standard of AS/NZS 1891.4 and, most importantly, its recent amendments. Early wording of this document allowed for anchors to be installed that held only 6 kN, assuming that they would be used only for work positioning. (That is, there would be no fall involved.) This has now changed and the new anchors must be able to withstand 15 kN: the force of a person falling 2 m (the length of most lanyards) plus the associated deployment lengths of the energy absorber and the person's height, as described earlier.

Note that, even then, the anchor point is designed for one person only. If two people are to be connected to the same anchor, then 21 kN of restraint is required.

In addition, anchor points generally only have a service life of 10 years. You must always check the load rating and installation date of any anchor system prior to using it.

Static lines and rails form another category of anchorage points, the difference being that you can attach to them and slide your hook or karabiner along the line or rail (Figure 1.60). These must be carefully engineered to AS/NZS 1891.2 standards, and are generally supplied and installed by a manufacturer as a complete system.

HOW TO

ANCHORAGE POINTS, STATIC LINES AND DOUBLE LANYARDS

There will be occasions when a work area will have either a static line running between multiple points of anchorage (Figure 1.60), or there will be no line and just multiple points of anchorage. Vertical ladders also replicate this situation. In such instances you will need a double lanyard system. In use, you must always keep one lanyard coupled to the line, anchor or ladder. The second lanyard comes into use when you need to move beyond a point where your hook or karabiner can't slide past. Or, when using the fixed vertical ladder without a static line or rail, you connect to every third or fourth rung, depending upon your reach – but always with one lanyard connected at all times (Figure 1.61).

Source: Shutterstock.com/Opsorman

FIGURE 1.61 Double lanyards

Risks and hazards of IFAS use

In covering the components of IFAS, we have considered the importance of correct fitting, the calculation of deployment distance, and the selection and application of lanyards and inertia reels. This leaves the vital issue of the risks and hazards of opting to use one of these systems in the first place.

In choosing to adopt any system intended to limit the potential for a fall, you must always ensure that what you are implementing is not introducing new, and potentially greater, risks. This applies to all control levels, not just the IFAS. Some systems will always have risks that cannot be completely eliminated and so must be acknowledged and worked around, through information, careful training and supervision. The IFAS approach, in particular, falls into this category.

The main risks associated with using an IFAS can be summarised as follows:

- over-confidence leading to an unnecessary fall
- trip and entanglement hazard from the lanyard, again leading to a fall
- failure to allow sufficient clear fall space - that is, the person hits the ground or another level before the system can fully operate
- swing or pendulum effect, leading to the person hitting a wall or other structure
- significant trauma or injury to the body caused by the harness from the shock of a fall
- poor fitting of harness, leading to greater injuries than those described above.

Understanding the trauma of suspension

When an individual is suspended on a rope or in a harness, even for a few minutes, they may go into shock. This is brought about when blood flow to and from the legs is significantly impaired, reducing the amount of blood available to the heart. This is a highly dangerous situation for the suspended person and can lead to them fainting, suffering permanent muscle or tissue damage and, in the worst instances, death.

Depending upon a number of factors (such as weight, nature and severity of restriction, general fitness and health, length of fall and/or associated injuries), a suspended person may become unconscious within minutes, or hang seemingly unaffected for a couple of hours (yet may succumb to complications later). Many of these factors will be unknown to the rescuer, so it is critical that timely and carefully considered rescue procedures are followed.

HOW TO

YOUR RECOVERY PLAN: ESSENTIAL PREPLANNING FOR USING AN IFAS

Get them down, give them fluids and lay them down – now!

It is important that an emergency recovery plan is in place whenever you use an IFAS. Without such a plan, and its immediate implementation, you will increase the chance, and/or severity, of injury to the suspended person.

From the outset, you should plan to lower the fallen person wherever possible. Lifting is extremely dangerous, as it increases the tension of the harness on the suspended person and hence increases the likelihood of injury.

Your recovery plan should include:

- immediate contact of emergency services (dial 000) – the fall itself is enough to warrant an emergency examination at a hospital
- advice to be given to the suspended person on how to reduce the chance/severity of going into shock (e.g. raising of legs, use of relief straps if available,

>>

movement that will reduce the pooling of blood in the legs)
- location of recovery equipment
- names and contact details of personnel on-site trained in recovery procedures
- the procedures for recovery: Down whenever practicable – this puts significantly less strain on the suspended person – raising a person is a dangerous last resort and must be done extremely slowly
- procedures for caring for a recovered person once they are on safe ground; that is, give fluids and place in recovery position unless there is some significant reason not to
- names and contact details of first-aid officers trained in treating suspended persons
- what the recovery equipment consists of; for example, an EWP or scissor lift that is always located on-site
- the people that are authorised to operate the recovery equipment
- a schedule that ensures an operator is on-site at all times while people are working at height.

Your plan presupposes that in using this equipment you will have someone on-site able to fulfil the roles mentioned above; that is, first-aid officers trained in dealing with suspended persons, and those trained in the recovery procedure (and any associated equipment). Given a timeline of minutes in some instances, it is critical that this be the case.

Finally, when using these systems, it is strongly recommended that relief steps (Figure 1.62) be fitted to the harness. Relief steps allow a worker to effectively 'stand' (Figure 1.63) in the harness and so take the pressure off the blood vessels around the groin and thigh areas. Even with the relief step in use, recovery of the suspended person needs to be undertaken as soon as safely practicable. **Your guiding rule: get them down and lay them down – now!**

FIGURE 1.62 Relief step package

FIGURE 1.63 Relief step in use

Source (images): Reproduced with permission of Honeywell Safety Products Pty Ltd, http://www.honeywellsafety.com.

COMPLETE WORKSHEET 3

Level 5 controls: ladders and administrative controls

This section covers two areas we often take for granted: ladders and administrative controls – the prior because we use them too regularly and have become used to considering them our first choice, and the latter because we often ignore them, considering paperwork boring and useless. However, we take both these areas for granted at our peril.

LEARNING TASK 1.6

1 **Circle 'True' or 'False'.**
A harness should be worn loosely so as to reduce the shock on the body should you fall.
True False

2 **Circle 'True' or 'False'.**
With a relief step fitted to a harness you will not need to visit the hospital if you fall and are suspended for any length of time.
True False

Ladders

And so we come finally to the access equipment at which most of us would ordinarily (but perhaps mistakenly) start: the humble **ladder**.

Unfortunately, ladders still show up in reports as a key killer of our workforce. For example, in the four years between 2016 and 2020, falls from heights resulted in 48 construction-related deaths across the country. Of these, 21 per cent were specifically related to falls from ladders (SWA, 2021). These numbers have not improved much on previous years given that between 2014 and 2018, 51 workers died from falls in construction accidents, only 18 per cent of which were due to ladder falls (SWA, 2021). Hence in percentage terms, fatal falls from ladders have actually increased – from 18 per cent to the current 21 per cent. Notably, in 2003 the percentage of fatalities attributable to ladders was only 7 per cent. We still do not seem to be learning very fast on this one!

Types of ladders

Ladders take a variety of forms and are used in numerous ways to undertake work at heights. The most familiar types are **step ladders** (Figure 1.64) and **single ladders** or **extension ladders** (Figure 1.65). These remain as the mainstay of workers in the construction industry (be it domestic housing or industrial), and will be found in just about any plumber's or builder's ute. They are also the cause of far too many construction-based injuries: this is because, used inappropriately, they are remarkably unstable.

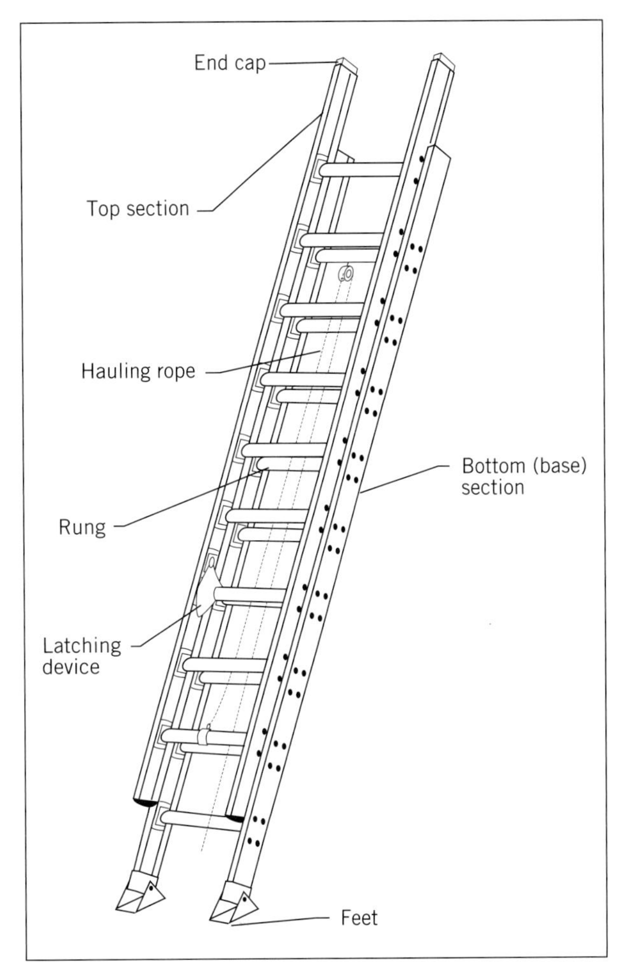

FIGURE 1.65 Extension ladder

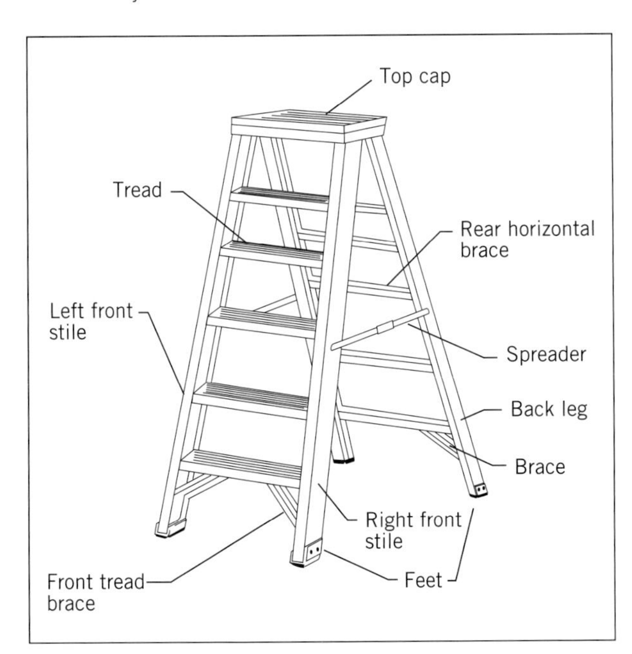

FIGURE 1.64 Step ladder

Ladders are made up of the following components: rungs (single or extension ladders) or steps (step ladders) and **stiles**. In addition, step ladders have a top plate and must have a spreader or locking system. Extension ladders will have a 'clutch' or locking mechanism, and may also have a rope and pulley system to aid raising and lowering. Some extension ladders also have adjustable locking feet for non-level ground.

All portable ladders must comply with the AS series AS/NZS 1892 Portable ladders, while fixed ladders are governed by AS 1657 Fixed platforms, walkways, stairways and ladders. In so doing, all ladders must be 'industrial' rated, not domestic or 'home-made' – that is, they must be rated to a minimum of 120 kg or greater.

Before using a ladder for any work, take a moment to consider the issues involved, remembering that ladders are your last resort, not your first. Go back over your hierarchy of control and determine if you can:

- complete the work appropriately using extensions on your tools
- bring it (or even part of it) down to the ground
- use a scaffold (mobile or otherwise)
- use a step platform
- construct fixed stairs or a solid work platform
- use an EWP
- incorporate a work position system.

In making your deliberations on the above, you should consider the following:

- nature of the task: the time, tools and materials required; stance and/or leverage and twisting actions needed (these put stresses on the ladder, causing it to slide, buckle or topple); any heavy lifting required; whether tools need two hands (electric drills of high torque, for example)
- context: Surroundings and conditions (noise, dust, water, debris, traffic, work being undertaken below, materials, and the like); what will the ladder be standing on; weather conditions if in an exposed position
- skills/training/knowledge level of the worker(s) using the ladder in respect of the task and working at heights
- accessibility: Despite the ladder, will you need to reach excessively outside the balance line of the ladder?
- does the task involve, or come close to, live electrical lines?
- what type or length of ladder would be required?
- when setting up ladders in or around walkways, doors or other trafficable areas, block, lock off or otherwise prohibit access until the work is complete.

We shall now consider the factors specific to the different ladder types.

ON-SITE

SAFETY NOTE

In considering all that, make sure you factor in the golden rule of ladders:

Always climb or descend facing the ladder,
never facing away from it.
This is a quick, short, sharp slide to disaster.
Also, wherever possible, work facing the ladder.

Step ladders

Much of the work undertaken on step ladders involves a potential fall of much less than 2 m. That said, people die from hitting their heads when just falling over, let alone falling 1 m while holding a 750 w drill with a spinning 38 mm spade bit in it. So:

- choose your ladder with care. Domestic ladders are not suitable for construction work (Industrial ladders are clearly marked, indicating a load rating of 120 kg or greater.)
- never (and that means not ever!) stand on, or work from, the last rung before the top plate, or the top plate itself, to undertake work
- always ensure the step ladder is fully expanded before standing upon it
- always ensure the ladder is positioned securely and vertically on stable ground
- if you must pack under the legs, do so using full-width timbers and never bricks or blocks, which can fracture suddenly under load
- if leaning the ladder against a wall (in the closed position), ensure the structure you are leaning it against is secure
- for large step ladders, when working near the top, have someone hold the base of the ladder.

Single or extension ladders

Invariably, the work undertaken from these ladders will have a potential fall of more than 2 m. In addition, the ladder design is such that *it is always reliant upon something else to hold it up*. Further, they are generally significantly taller than the individual who is carrying, positioning or repositioning them. This means that it is very easy to hit something above, behind or even in front of you. 'Look up and live' is the catch cry here. This alludes to the potential for striking live power sources, such as power lines from the street to a house, or light fittings and the like. So, in choosing, installing and using an extension or single ladder, the procedure set out in the following box is recommended practice.

HOW TO

PROCEDURE FOR LADDER USAGE

1. Never use a ladder with missing or damaged components.
2. Use non-conductive ladders (fibreglass or timber without metal stile-reinforcing wires) for work on or near electrical supplies.
3. Ensure that the structure you are leaning the ladder against is sound and capable of withstanding the potential loads of ladder, workers, tools and materials.
4. Ensure the ladder is on a solid, level surface, and that both feet are in contact.
5. Install the ladder at an angle of 1:4.
6. Ensure that both stiles are in contact at the top.
7. Ensure that the extension ladder lock or 'clutch' is fully engaged before climbing.
8. Ensure that the top of the ladder extends at least 1 m (or three rungs) past a landing or platform when being used as access.
9. Never stand higher than 900 mm from the top rung of the ladder (or four rungs from the top).
10. Secure the bottom of the ladder so that it cannot slide, or have someone hold the base of the ladder.
11. When climbing or descending, have three points of contact with the ladder at all times.
12. Secure the top of the ladder so that it cannot slide left or right. Do so by tying off or using a ladder bracket. (See Figure 1.66.)

Best practice is to have someone hold the ladder at the base until you have secured the top.

Use pole straps when working against columns, poles or trees.

>>

>> **Additional safety notes**

- When working from the ladder, where possible ensure your hips are within the stiles. Leaning past this tends to push the ladder sideways.
- Never allow work to take place beneath a ladder.
- Do not climb from ladder to ladder.
- Never use extension ladders in a horizontal position (e.g. as a 'bridge').

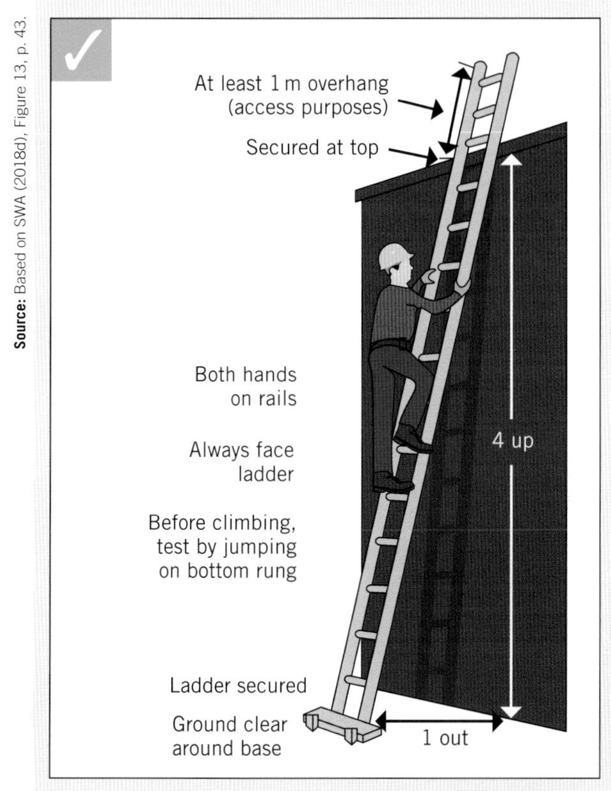

Source: Based on SWA (2018d), Figure 13, p. 43.

FIGURE 1.66 Recommended practice for ladder usage

Fixed ladders

On some projects, you may be required to access a work area on an existing structure using a ladder that has been permanently fixed in place (Figure 1.67). Such ladders, known as 'fixed ladders', are often quite steep, over 75 degrees to the horizontal, and may have a cage installed around them. Permanent rooftop walkways may also be provided on a building that you are renovating, repairing or installing new elements to.

The installation of such features is governed by AS 1657 Fixed platforms, walkways, stairways and ladders – Design, construction and installation. This standard also recommends that safe systems of work be adopted, which may include anchorage lines, rails and fall arrest systems.

It is vitally important that you make yourself aware of the particular system required for the type of access you are confronted with in each instance. This will be included in your SWMS and it should outline clearly the steps you will take. This will most likely include the adoption of an IFAS – wearing a harness and using a double lanyards – to ensure connection at all times as you transit along the walkway or transition from the ladder to the roof top.

FIGURE 1.67 Vertical access ladder

Note: Ladder cages can inhibit rescue attempts in some instances (they frequently do not stop a fall …). A rescue procedure, designed specifically for the given access system, should be developed as part of your SWMS.

Maintenance of ladders

As stated earlier, no ladder should be used if any component is missing or damaged. In addition, there are certain things that you can do, and things you should not do, to ensure that ladders remain serviceable:

- Never paint a ladder, be it timber, fibreglass or otherwise. Paint can disguise or hide points of failure, cracks or important safety information. It can also make rungs or steps slippery.
- Store ladders carefully. When storing for transport, such as on the roof of a vehicle or in the back of a ute or truck, ensure that the ladder is able to lie straight. The ladder should also be supported (if on the roof) in a manner that does not induce significant flex (such as hanging out the back of the ute, being held up at the very ends of the ladder only, or laying over something at its midpoint).
- Check that the clutch, guides and, where fitted, ropes and pulleys are in good order and working.
- Check for wear of any moving parts of step ladders, such as hinges and spreaders.
- Check rivets on braces for wear or tear, and that they are not missing.

- Ensure that levelling feet on extension ladders are functioning properly and cannot slip from the locked position.
- Ensure that any opaque materials (such as paint or glue spills) are cleaned from the ladder.
- Ensure that all safety alert signage on the ladder is still legible.

Using ladders as scaffold: ladder bracket scaffolds

The use of ladder brackets (Figure 1.68) to create a scaffold is not considered acceptable for general construction work by the National Falls COP, yet they are not prohibited except in Queensland. As of 2016, WorkSafe Queensland made ladder brackets illegal on all construction sites, and discouraged their use in any other workplace. They are, however, *never* to be used where a potential fall may be greater than 2 m, because they can only carry one plank, which is narrower than the required minimum of 450 mm for work at that height.

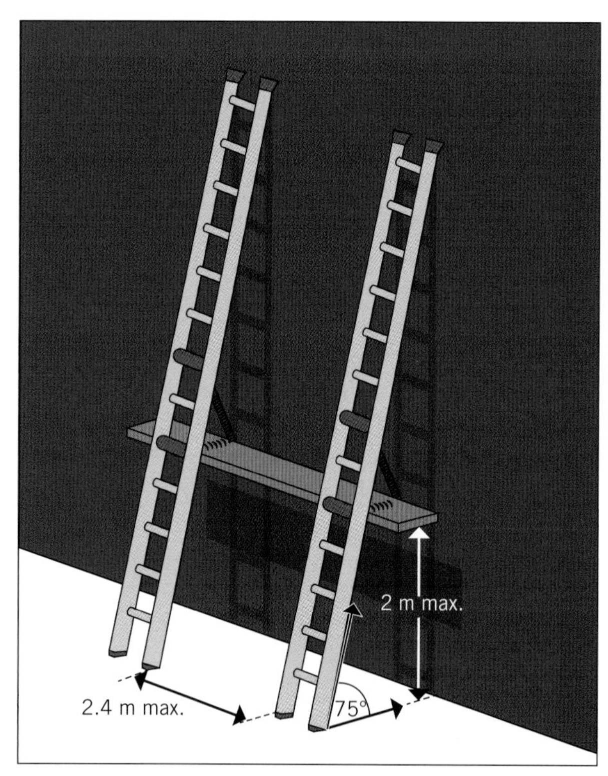

FIGURE 1.68 Ladder bracket scaffolding

If it is absolutely necessary to use this system, then ensure the following:

- the plank is *never* installed at a height greater than 2 m from the surface below
- the ladders are correctly positioned at 1:4, and are securely fixed top and bottom
- the ladder brackets are correctly fitted to both ladders and the plank
- only one person works on the system at any one time. This also means that no one can climb either ladder while someone is working from the plank
- barricades and signage are set up to stop people or vehicles from passing close to, or making contact with, the system
- the plank is a genuine **scaffold plank** and is in good order
- the distance between ladders does not exceed the stated SWL of the plank (the maximum distance the plank can span) or 2.4 m maximum (whichever is less)
- the ends of the plank extend past the brackets by *at least* 150 mm and not more than 250 mm
- the system is 'under slung', as shown in Figure 1.68. The plank is *never* allowed to be positioned across the front of the ladder.

Having noted the above points, keep in mind that there are many more effective methods for working at heights than ladder **bracket scaffolding**.

Administrative controls

Administrative controls have, to some degree, already been discussed in previous sections: they include SWMS documents (and the like), as well as documented procedural information such as emergency procedures and approved methods for conducting tasks. Administrative controls are the backstop to all the previously discussed control levels.

Falling back on administrative controls alone, however, is a last-ditch effort. It is an attempt to do 'something' to reduce the exposure of workers to potential falls. This 'fall-back' position should only be adopted when it is absolutely impossible to implement any of the higher control levels (and you have gone over them at least three times before coming to this point). The key limitation of administrative controls alone is that they are highly reliant upon competent and vigilant supervision.

Administrative controls may be used effectively to:

- stipulate specific work procedures that must be followed
- limit the time workers may be involved in a particular task or activity
- stipulate the use of specific equipment
- provide maximum and minimum numbers of workers for particular tasks
- document supervision, and give guidance on how that supervision must be conducted (time frames and the like)
- document permit systems for limiting access and monitoring egress (ensuring that workers have left a site, or section of a site)
- stipulate and help enforce 'no go' areas
- monitor vehicular traffic.

For administrative controls to be effective, constant supervision and completion of any documentation requirement is absolutely essential. In addition, they must be audited and reviewed on a frequent basis to ensure this adherence, and be improved where necessary.

LEARNING TASK 1.7

1. **Circle 'True' or 'False'.**
   The key limitation of administrative controls alone is that they are highly reliant upon competent and vigilant supervision.
   True False
2. **Ladder bracket scaffold may only be used if the distance from the ground to the working platform is equal to or less than:**
   a 2 m
   b 3 m
   c 4 m
   d 5 m

Best practice for common tasks

Applying all this information can be challenging. Guidelines for undertaking some of the more common tasks that involve working at heights follow. Where possible, you should add to this advice any information you can obtain from manufacturers, suppliers, or the relevant codes of practice or Australian Standards. (See 'References and further reading' at the end of this chapter.)

Laying floors to upper storeys

The following tasks may involve a number of tools being used that can add to the risks of working at height. For example, using an air compressor, hose and nailing gun can present a number of hazards including tripping over the hose, an unexpected release of high pressure air from the hose if punctured and potential manual handling injuries.

Wherever possible it is useful to avoid tools such as air compressors and the associated hoses when working at height. Upgrading to a battery-operated nailer is usually more efficient and eliminates several hazards.

Various power tools introduce similar hazards, such as a power lead running to a height often gets tangled and can cause a tripping hazard or become an electrocution risk. Like the battery-operated nailer, any power tools that are needed to complete construction work at height can be battery operated which eliminates these hazards entirely.

The preferred method of construction – from both time and safety, as well as energy efficiency, perspectives – is to 'sheet out' a floor before establishing the walls. This method is best practice for ground-level stump and bearer construction where potential falls are generally less than 2 m, as well as being the preferred approach for upper-level floors where falls may be significantly greater. The procedure is set out in the following box.

HOW TO

PROCEDURE FOR LAYING FLOORS TO UPPER STOREYS

1. Install edge protection (Figure 1.69).
2. Lay the first sheet(s) from either an exterior or interior scaffold or platform. (This generally means that the initial potential fall height is less than 2 m.)
3. Once the first sheets are in place, apply edge protection.
4. Apply the rest of the sheets by working off each of those laid previously.

Caution: You still have a potential fall of greater than 2 m 'in front' of you as the floor progresses (Figure 1.70). This is particularly evident when applying adhesives. A static line running the length of the building and an adjustable travel restraint system without a shock absorber (its extended distance would mean you would hit the floor below) may be appropriate here.

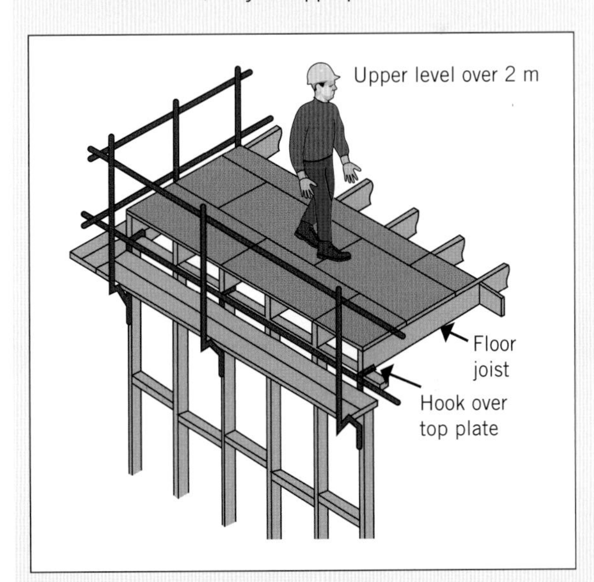

FIGURE 1.69 Edge protection

Source: SWA (2018d), Figure 16, p. 48.

FIGURE 1.70 Gluing sheets on upper floors

Source: Alamy Stock Photo/David R. Frazier Photolibrary, Inc.

Framing upper roofs

Standing wall frames always come with the risk of either pushing the wall itself off the floor edge or stepping off the edge yourself. Upper-level wall framing comes with the added risk of a fall of significantly more than 2 m. An IFAS is not an option, as the fall is still generally less than the 6.5 m required for deployment. The following box sets out what is regarded as the best practice for this task.

Figure 1.71 shows the correct approach for standing upper-level wall frames. (It is not dissimilar to the approach used by most builders for standing lower-level walls.)

Note: Perimeter guardrails have been deleted in Figure 1.71 for clarity.

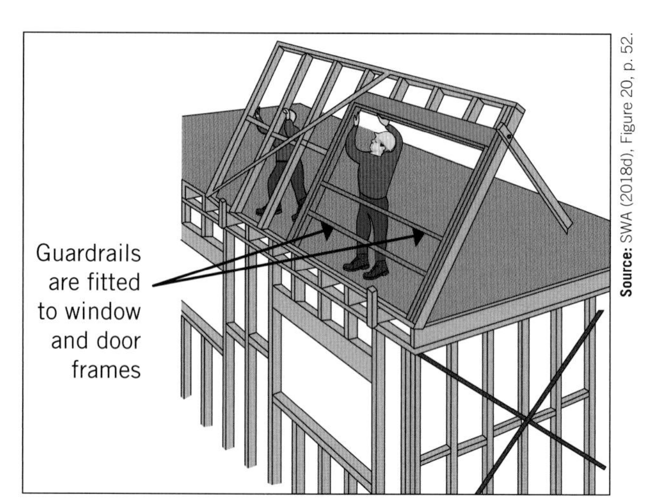

FIGURE 1.71 Standing upper-level wall frames

HOW TO

PROCEDURE FOR FRAMING UPPER ROOFS

1 Establish perimeter protection and guardrail all voids.
2 Fix temporary stops to the outside of the wall plates.
3 Position the bottom plate of the wall in line with the floor edge and up against stops. (Avoid letting the wall 'slide' into stops, as they may sheer off.)
4 Attach a rope to the top plate as a restraint against the wall going too far past vertical and toppling off the building (not visible in Figure 1.71).
5 Attach a brace to a convenient stud, as shown in Figure 1.72.
6 Fit guardrails over any window or door openings that have sill trimmers lower than 1 m.
7 Stand the wall and fix the lower end of the **brace**.

Note: Without a perimeter railing, you could do this with travel restraint systems in place. However, such systems may be more awkward and could lead to other hazards that would need to be carefully considered.

HOW TO

PROCEDURE FOR ASSEMBLING ROOF TRUSSES

(The following assumes that walls have appropriate temporary bracing and have been marked out for truss positions.)

1 Whenever possible, have trusses placed by crane directly from the delivery truck on to the top wall plates.
2 Ensure that trusses are located and stacked in reverse order to their use. This reduces handling and unnecessary strain on workers.
3 Wherever possible, establish a working platform with a minimum width of 450 mm (a trestle scaffold is acceptable for this) on the inside of the building, rather than the outside. This often reduces the potential fall height to less than 2 m and uses the wall as a handrail on one side at least. Workers must not stand on the external top plates at any time.
4 Place rails across openings (windows or doors) through which a person may fall. This is critical if the potential fall is greater than 2 m.
5 Planks may be positioned across internal top plates, and workers may work off these plates provided that:
   - they remain 1.5 m from the external walls
   - they cannot fall into a stairwell or void
   - planks do not exceed their rated spans.

For trusses at 600 mm centres:

6 The standard assembly procedure allows that once the first truss is stood and adequately braced, the other trusses may be temporarily braced from it and each other (Figure 1.72). To do this, you may stand on and walk the bottom cords, provided that:

>>

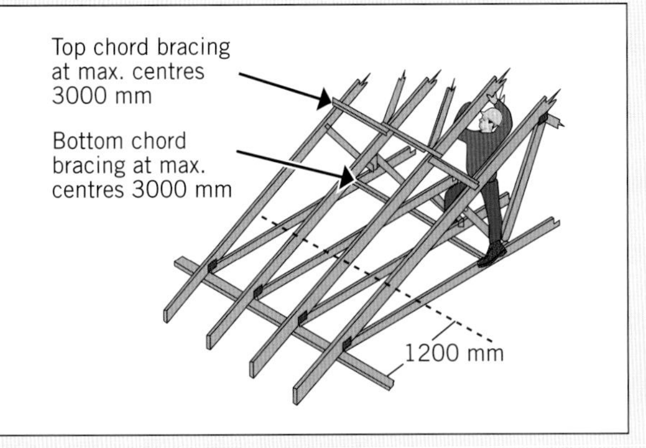

Source: Based on SWA (2018d), Figure 22, p. 55.

FIGURE 1.72 Temporary ties between trusses

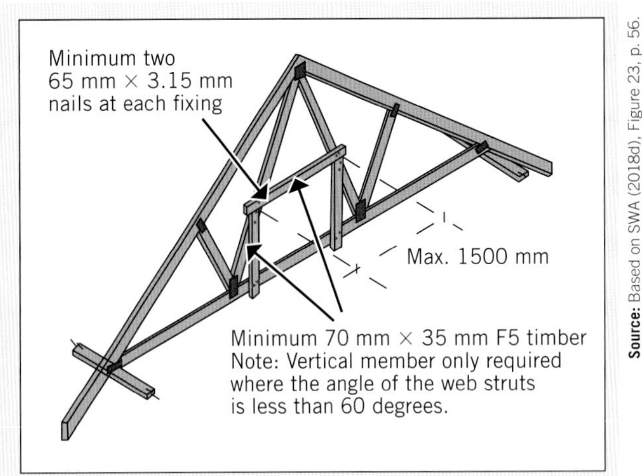

Source: Based on SWA (2018d), Figure 23, p. 56.

FIGURE 1.73 Waling plates

- trusses are no more than 600 mm apart and are adequately braced
- suitable footwear is worn (good grip)
- the work is supervised by a competent person
- the competent person visually checks the bottom cords before they are stood upon.

7 When the apex is out of reach, 'waling plates' (Figures 1.73 and 1.74) may be installed to every fourth truss pair.

8 Once all trusses are up, a single run of sheet flooring 900 mm wide with toe boards may be run down the centre of the bottom cords. Lateral cord ties can be fixed at 900 to 1100 mm in height to act as handrailing.

This improves safe access for final bracing and future fit-out of electrical and plumbing lines.

For trusses at greater than 600 mm centres:

9 These trusses provide a greater challenge, as falls are harder to prevent if 'walking' the truss bottom cords. In these cases, use one or combinations of the following:
- mobile scaffold
- safety netting
- EWPs

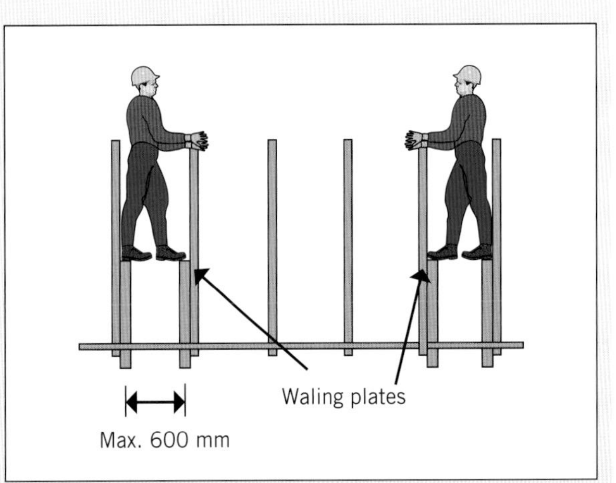

Source: Based on SWA (2018d), Figure 24, p. 57.

FIGURE 1.74 Standing on waling plates

- planks across the bottom cords and top plates of the internal walls. Planks must produce a platform with a minimum clear walking width of 450 mm.

Conventional, pitched or 'stick' roofs

The initial approach for this form of roof is similar to that outlined for trusses – that is, work within the external walls wherever practicable; otherwise, create a handrailed scaffold platform around the external walls. The procedure is set out in the following 'How to' box.

HOW TO

PROCEDURE FOR CONVENTIONAL, PITCHED OR 'STICK' ROOFS

Following the ceiling joists being installed:

1 Use sheet flooring or planks over these members to give yourself a safe work area.

2 When a ridge must be installed at a height out of reach of the raised work surface, the sheet flooring must be extended so as to effectively produce a 'catch platform'. This means that it must extend a minimum of 2 m beyond the potential fall area. This would allow an additional maximum height of 1 m to be achieved by means of plastering stools and/or planks only. For higher than this, see below.

There are three alternatives to the above.

1 Use an EWP.
2 Install scaffolding direct from ground level.
3 Construct the entire roof at ground level on a welded steel subframe and crane-lift it into position.

Battens to roofs

Limitations apply to the installation of battens without the use of EWPs, safety netting or internal catch platforms. These limitations are:

- trusses at 600 mm centres: battens at maximum 900 mm spacing
- trusses at 900 mm centres: battens at maximum 450 mm spacing
- trusses at 1200 mm centres: battens at maximum 450 mm spacing and installed sequentially (from the bottom up).

Irrespective of the truss or batten spacings involved, the procedure set out in the following box is recommended (assuming a perimeter guardrail is in place or external catch/work platforms are installed).

HOW TO

PROCEDURE FOR INSTALLING BATTENS TO ROOFS

1. Set out batten location on trusses prior to standing trusses. This is more easily done by marking them as a single set while stacked on the plates.
2. Partially drive in nails at each mark.
3. With trusses in position, locate first batten (lowest, and so closest to wall) and fix.
4. Position and fix the remaining battens sequentially, using the nails as a means of temporary location.
5. When batten spacings are within the above-listed limitations, walking the battens is permitted.

Installing safety mesh

Roofing plumbers will often need to install safety mesh as the preferred means of fall protection on large roofs, sheds and commercial buildings. When securely fixed, safety mesh prevents falls between rafters, purlins and/or portal frames. Given the size of some roofs, it is the ideal means of prevention for the following reasons:

- cost-effectiveness
- ease of installation
- permanent installation
- remains as a fall prevention measure for future roof repairs, maintenance or post-construction installations (e.g. air-conditioner units).

Safety mesh prevents falls only internal to the structure, and so must be used in conjunction with other measures such as:

- perimeter guardrails
- exterior scaffolding (work or catch platforms)
- IFAS.

All safety mesh installations must comply with AS/NZS 4389 Safety mesh. This code requires the mesh to be made of minimum 450 MPa tensile strength wires of 2 mm diameter or greater, with a grid of not more than 150 mm × 300 mm between strands.

HOW TO

PROCEDURE FOR INSTALLING SAFETY MESH

The following procedure assumes that appropriate edge protection/scaffold/catch platforms are in place. It also assumes competent installers and/or supervisors to ensure that all fixings, materials and arrangement of mesh complies with AS/NZS 4389.

1. Measure or calculate the total roof length to determine the length of mesh required to clear cover from gutter to gutter.
2. Cut the mesh to length while on the ground, then roll up and position on one side of the building.
3. Attach a rope to one end of the mesh and cast the rope to the other side of the building.
4. Pull the sheeting over, as shown in Figure 1.75. (*Do not* walk on open purlins to pull mesh into place.)

When purlin spacing is greater than 1.7 m, the usual 150 mm side overlap is not considered sufficient. In these cases, every second square needs to be linked by means of a 2 mm staple. These staples must be fixed by a competent person from underneath the roof. This generally means using an appropriate EWP. It must never be attempted from above.

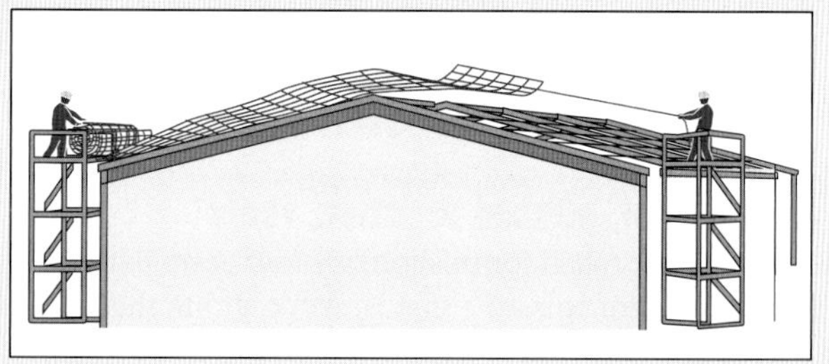

FIGURE 1.75 Installing safety mesh

Source: WorkSafe Victoria (2019), Figure 26, p. 57.

>>

>>

Walking a meshed roof

Safety mesh is a means of fall protection only. It is not to be deliberately walked or lain upon, particularly when carrying loads such as roof sheets or tools.

When walking a meshed roof for the purposes of sheeting, observe the following:

- Walk upon the top cords of portal frames, as shown in Figure 1.76.
- When necessary, walk the edge of purlins with great care (Figure 1.77).
- Only walk boxed gutters when you have advice that they have been constructed with adequate support (Figure 1.78).
- Have existing or old box gutters inspected by a competent person before walking them.

FIGURE 1.77 Walk the edge of the purlins only when necessary

FIGURE 1.76 Walk on the top cords when walking a meshed roof

FIGURE 1.78 Walk boxed gutters only if they have been constructed adequately

Fascia, gutter and roof sheet installation

This work should be done using perimeter scaffold or a catch platform. The exact form that each of these will take is dependent upon the actual pitch of the roof (as outlined previously under the heading 'Edge and perimeter protection'). Generally, fascia and gutter are fixed first, followed by the sarking and roof sheeting.

The procedure from this point on is dependent upon a number of factors:

- pitch of roof
- spacing of rafters or trusses
- spacing of battens or purlins.

When the roof pitch exceeds 35 degrees, it is considered an inappropriate surface to stand, walk or work upon. In these instances, as outlined previously in 'Edge and perimeter protection', at least two of the following options must be in place (preferably including a scaffold platform as one option):

- work positioning system
- roof ladder
- scaffold platform.

HOW TO

PROCEDURE FOR FASCIA, GUTTER AND ROOF SHEET INSTALLATION

For a roof under 25 degrees, and where the batten/rafter/truss spacings allow, the following procedure is considered best practice.

1. Measure or calculate the overall length of the roof surface.
2. Cut sarking to length, re-roll, and position on roof or working platform.
3. Roll out the first sheet of sarking by walking the battens and affixing as required.
4. Position and provisionally fix the first line of sheeting.*
5. Now lay the rest of the sarking and sheeting by working off the previously laid sheets.

* *Note:* It is preferred practice to 'fix off' all the roof sheets as one, using string lines once they are in position. However, there is a danger with using this approach in high winds. If the full roof is not going to be completed within the one day, sheets must be final fixed and not left overnight with provisional fixings only. Even so, provisional fixing should be sufficient to withstand any sudden wind gusts that might occur prior to final fixing.

Clear the work area

Once the work at height has been completed, it is important to clean the work area. If this task is not completed before the fall prevention equipment is removed it will cause issues later on. For example, a plumber that has completed a skylight installation forgets to clean away the swarf, rivet stems and pieces of sheet metal that have been left. To access the roof area again the fall prevention system will need to be reinstated, adding expense and time to the job.

Ensure the roof area is swept clean and always recycle materials where possible before the fall prevention system is removed.

Exiting the work area

When exiting from the work area it is important to ensure all tools and materials are removed and checked in accordance with worksite procedures, and safety and environmental requirements.

Fall prevention systems must be dismantled and removed from site at the completion of the work. It is important to follow the manufacturer's instructions when dismantling equipment and only trained people should complete this work. Where a license is required to construct; for example, scaffolding over 4 m, a licensed person should also be used to dismantle the scaffolding.

Any damage that is observed during the dismantling process to equipment or tools should be immediately reported to the supervisor and the equipment or tools should be tagged out.

Scaffolding, work platforms and other structural equipment should be frequently inspected for signs of damage. This can be achieved through a simple visual inspection, carrying out checks to ensure none of the scaffolding components have been removed or have been damaged through the course of the construction work taking place.

If the damage is observed in a scaffolding or similar the area at height should be not be accessed until that fall prevention control is repaired and reinstated.

GREEN TIP

Avoiding the hazard of working at heights by building a platform floor on joists and then constructing framework on that platform aids doing the job efficiently which helps the environment.

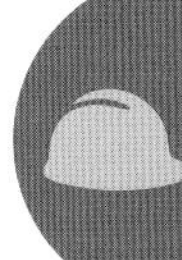

Falling from any height can be dangerous. The greater the height, the greater the likely injuries. However, the surface that you might fall onto is critical in your planning. A fall from a low height onto protruding reinforcement bars can be life changing.

COMPLETE WORKSHEET 4

SUMMARY

The purpose of this chapter has been to assist you in developing the necessary knowledge and skills to safely undertake construction work at heights. In particular, the chapter guides you through the statutes and regulations applicable to your specific state or territory context. While there is a national model, the regulations and the language used vary from state to state, OHS and WHS being the key examples.

In completing the chapter and its associated worksheets, you should be aware, as a minimum, of the following key points:

- Working at heights is not limited to working above 2 m.
- Once you are working at heights greater than 2 m, you are required by legislation to carry out more stringent controls.
- Identification of the hazards, risks and potential choices of control must be derived through the use of some form of SWMS/JSA/SWP document.
- In all states and territories, SWMS are mandatory for all high risk construction work, which includes working at heights.
- You should base your choice of control on a five-level hierarchy, with the first being always to explore avenues for not having to work at heights at all.
- Ladders should always be your last resort, not first thing you think to use.
- If your control includes the use of an IFAS, then you must have a recovery plan and ensure that the underlying premise of this plan is to get the suspended person down, and lay them down, without delay.

Understanding the benefits, and particularly the limitations, of your various control options is a critical aspect in development towards you and your team safely working at heights. Remember that you have five levels to work through, and that while each level has its relevance, as you work through levels 1 through to 5 the associated risks tend to increase. When undertaking common tasks, it is therefore wise to follow existing best practice in each case. Some examples of best practice were described towards the end of this chapter.

Always keep in mind that working at heights is a high-risk activity. Your decisions can, and often will be, life or death. Pause, consider and make those decisions wisely.

REFERENCES AND FURTHER READING

Text

Australian Building Codes Board (2022), *National Construction Code (NCC), Volume Two*, Commonwealth of Australia, Canberra.

Australian Safety and Compensation Council (2006), *National Standard for Licensing Persons Performing High Risk Work*, Australian Safety and Compensation Council, Canberra, **https://www.safeworkaustralia.gov.au/system/files/documents/1702/nationalstandard_licensingpersonsperforming highriskwork_2006_pdf.pdf**

Safe Work Australia (2011), Interpretive guideline – Model Work Health and Safety Act – the meaning of reasonably practicable, **https://www.safeworkaustralia.gov.au/system/files/documents/1702/interpretive_guideline_-_reasonably_practicable.pdf**

Safe Work Australia (2018a), *Construction Work: Code of Practice*, Commonwealth of Australia, Canberra, **https://www.safeworkaustralia.gov.au/sites/default/files/2022-10/Model%20Code%20of%20Practice%20-%20Construction%20Work%20-%2021102022%20.pdf**

Safe Work Australia (2018b), *Excavation Work: Code of Practice*, Commonwealth of Australia, Canberra, **https://www.safeworkaustralia.gov.au/sites/default/files/2022-10/Model%20Code%20of%20Practice%20-%20Excavation%20Work%20-%2021102022.pdf**

Safe Work Australia (2018c), *Managing the Risk of Falls at Workplaces: Code of Practice*, Commonwealth of Australia, Canberra, **https://www.safeworkaustralia.gov.au/sites/default/files/2022-10/Model%20Code%20of%20Practice%20-%20Managing%20the%20Risk%20of%20Falls%20at%20Workplaces%2021102022_0.pdf**

Safe Work Australia (2018d), *Managing the Risk of Falls in Housing Construction: Code of Practice*, Commonwealth of Australia, Canberra, **https://www.safeworkaustralia.gov.au/system/files/documents/1810/model-cop-preventing-falls-in-housing-construction.pdf**

Safe Work Australia (2021), Work-related traumatic injury fatalities Australia 2020, Safe Work Australia, Canberra, **https://www.safeworkaustralia.gov.au/sites/default/files/2021-11/Work-related%20traumatic%20injury%20fatalities%20Australia%202020.pdf**

Safe Work Australia (2022a), *Model Work Health and Safety Act*, Commonwealth of Australia, Canberra, **https://www.safeworkaustralia.gov.au/doc/model-work-health-and-safety-act**

Safe Work Australia (2022b), *Model Work Health and Safety Regulations*, Commonwealth of Australia, Canberra, **https://www.safeworkaustralia.gov.au/doc/model-whs-regulations**

South Western Sydney Institute of TAFE NSW (2017), *Basic Building and Construction Skills*, 5th edn, Cengage Learning Australia, Melbourne.

Relevant regulatory authorities and legislation

Australian Federal Government

Authority: Comcare
Level 4 (Reception)
14 Moore Street, Canberra ACT 2600
Phone: (02) 6275 0000
Workers' Compensation: 1300 366 979
WHS Hotline: 1800 642 770
Website: **http://www.comcare.gov.au**
Sources: ohs.help@comcare.gov.au

Australian Capital Territory
Authority: WorkSafe ACT
Phone: (02) 6207 3000
Website: **https://www.accesscanberra.act.gov.au/app/home**
Work Health and Safety Act 2011;
http://www.legislation.act.gov.au

New South Wales
Authority: SafeWork NSW
Phone: 13 10 50
Website: **http://www.safework.nsw.gov.au**
Work Health and Safety Act 2011
Work Health and Safety Regulation 2011 (Amdt 2019);
http://www.legislation.act.gov.au

Northern Territory
Authority: NTWorkSafe
Phone: 1800 019 115
Website: **http://www.worksafe.nt.gov.au**
Work Health and Safety Act 2011 (Amdt 2019)
Work Health and Safety Regulations 2011 (Amdt 2019);
https://nt.gov.au

Queensland
Authority: WorkCover Queensland
Phone: 1300 362 128
Website: **http://www.worksafe.qld.gov.au**
Work Health and Safety Act 2011
Workplace Health and Safety Regulations 2011 (Amdt 2019)
Workplace Health and Safety Codes of Practice;
https://www.legislation.qld.gov.au

South Australia
Authority: SafeWork SA
Phone: 1300 365 255
Phone: (08) 8303 0400
Website: **http://www.safework.sa.gov.au**
Work Health and Safety Act 2012
Work Health and Safety Regulations 2012
Workplace Health and Safety Codes of Practice;
http://www.parliament.sa.gov.au

Tasmania
Authority: WorkSafe Tasmania
Phone: 1300 366 322
Phone: (03) 6166 4600 (outside Tasmaina)
Website: **http://www.worksafe.tas.gov.au**
Work Health and Safety Act 2012 (Amdt 2015)
Work Health and Safety Regulations 2012;
http://www.thelaw.tas.gov.au

Victoria
Authority: WorkSafe Victoria
Phone: 1800 136 089
Phone: (03) 9641 1444
Website: **http://www.worksafe.vic.gov.au**
Occupational Health and Safety Act 2004;
Occupational Health and Safety Regulations 2017;
https://www.legislation.vic.gov.au

Western Australia
Authority: Department of Consumer and Employment Protection (Worksafe WA)
Phone: 1300 307 877
Website: **http://www.commerce.wa.gov.au/WorkSafe**
Occupational Safety and Health Act 1984;
http://www.slp.wa.gov.au

(*Note:* It is the intent of the Western Australian government to adopt a version of the WHS Act and Regulations and at the time of writing it was still being considered by the WA government.)

Relevant Australian Standards

Fall-arrest equipment	AS/NZS 1891.1 Industrial fall-arrest systems and devices – Part 1: Safety belts and harnesses
	AS/NZS 1891.2 Industrial fall-arrest systems and devices – Part 2: Horizontal lifeline and rail systems
	AS/NZS 1891.3 Industrial fall-arrest systems and devices – Part 3: Fall arrest devices
	AS/NZS 1891.4 Industrial fall-arrest systems and devices – Part 4: Selection, use and maintenance
Rope access	AS/NZS 4488.1 Industrial rope access systems – Specifications
	AS/NZS 4488.2 Industrial rope access systems – Selection, use and maintenance
Portable and fixed ladders	AS/NZS 1892.1 Portable ladders – Part 1: Metal
	AS/NZS 1892.2 Portable ladders – Part 2: Timber
	AS/NZS 1892.3 Portable ladders – Part 3: Reinforced plastic
	AS/NZS 1892.5 Portable ladders – Part 5: Selection, safe use and care
	AS 1657 Fixed platforms, walkways, stairways and ladders – Design, construction and installation

>>

>>

Scaffolding	AS/NZS 1576.1 Scaffolding – General requirements
	AS/NZS 1576.2 Scaffolding – Part 2: Couplers and accessories
	AS/NZS 1576.3 Scaffolding – Part 3: Prefabricated and tube-and-coupler scaffolding
	AS 1576.4 Scaffolding – Part 4: Suspended scaffolding
	AS/NZS 1576.5 Scaffolding – Part 5: Prefabricated splitheads and trestles
	AS/NZS 1576.6 Scaffolding – Metal tube-and-coupler scaffolding – Deemed to comply with AS/NZS 1576.3
	AS 1577 Scaffold planks
	AS/NZS 4576 Guidelines for scaffolding
Platforms, walkways and stairs	AS 1657 Fixed platforms, walkways, stairways and ladders – Design, construction and installation
Edge protection	AS/NZS 4994.1 Temporary edge protection – General requirements
	AS/NZS 4994.2 Temporary edge protection – Roof edge protection – Installation and dismantling
	AS/NZS 4994.3 Temporary edge protection – Installation and dismantling for edges other than roof edges
Safety mesh	AS/NZS 4389 Safety mesh
EWPs and cranes	AS 2550.1 Cranes, hoists and winches – Safe use – General requirements
	AS 2550.10 Cranes, hoists and winches – Safe use – Mobile elevating work platforms
	AS 2550.16 Cranes – Safe use – Mast climbing work platforms
	AS 1418.10 Cranes, hoists and winches (design)
	AS 1418.17 Cranes (including hoists and winches) – Design and construction of workboxes

GET IT RIGHT

FIGURE 1.79 Examples of unsafe ladder use

A range of inappropriate ladder uses are described in Figure 1.79. Using the information found in Chapter 1 and associated references, outline methods for working at heights that would meet both the needs of the situations described and the safety requirements demanded in the contemporary Australian workplace.

WORKSHEET 1

To be completed by teachers

Student competent ☐

Student not yet competent ☐

Student name: ______________________________

Enrolment year: ______________________________

Class code: ______________________________

Competency name/Number: ______________________________

CPCCCM2012: Work safely at heights – specifically, the element 'Identify task requirements'

Also:

CPCCCM2010: Work safely on scaffolding higher than two metres – specifically, the element 'Identify work area requirements'

CPCCCM2010B: Work safely at heights – specifically, the element 'Identify work area requirements'

CPCPCM2055A: Work safely on roofs – specifically, the element 'Identify work safety requirements'

Task: Answer the following questions.

1 Who is responsible for enforcing WHS/OHS regulations in your state or territory?

____

____

____

2 List the three levels of regulation provided under the Model WHS Act.

1 ______________________________

2 ______________________________

3 ______________________________

3 Which of the above three levels are mandatory?

____

____

____

4 List the five steps used in creating a JSA, SWMS or SWP document.

1 ______________________________

2 ______________________________

3 ______________________________

4 ______________________________

5 ______________________________

5 Circle 'True' or 'False'.

You decide to develop your own working at heights procedure, rather than following the procedure that is outlined in the code of practice or guidance material. This action is allowed under the state regulations and will not automatically lead to a fine if 'found out'. True False

6 Aside from your fellow workers and employer, list who else you should speak to when working through your JSA/SWMS/SWP document.

____

____

____

____

____

7 Circle 'True' or 'False'.

A SWMS document is mandatory for all high risk construction work, whereas a JSA or SWP is not. True False

8 What is the definition offered for a 'competent person'?

____

____

9 Circle 'True' or 'False'.

Specific fall protection is required in any situation when the potential fall is greater than 2.0 m. True False

10 List the five levels of the hierarchy of control.

1 ______________________________________

2 ______________________________________

3 ______________________________________

4 ______________________________________

5 ______________________________________

11 Ladders are considered under which level of control?

____

12 List six key points you will need to consider when determining the task requirements, work area and access.

1 ______________________________

2 ______________________________

3 ______________________________

4 ______________________________

5 ______________________________

6 ______________________________

13 Of all the controls offered, which is always the preferred option? Why?

Option: ______________________________

____

Why: ______________________________

14 You have been asked to access a roof to install a small piece of ridge capping. The roof has a travel restraint system in place that can be used to reduce the risk of falling which your employer has asked you to use. To utilise this system in accordance with the code of practice 'Managing the risk of falls in housing construction', what two pieces of fall arrest equipment will you need while accessing the roof and what tools and materials will be required to complete the work?

____

____

____

____

WORKSHEET 2

To be completed by teachers	
Student competent	☐
Student not yet competent	☐

Student name: ______________________________

Enrolment year: ______________________________

Class code: ______________________________

Competency name/Number: ______________________________

CPCCCM2012: Work safely at heights – specifically, the element 'Access work area'

Also:

CPCCCM2010B: Work safely at heights – specifically, the element 'Access work area'

CPCCCM2010: Work safely on scaffolding higher than two metres – specifically, the element 'Access work area'

CPCPCM2055A: Work safely on roofs – specifically, the element 'Prepare for work'

Task: Answer the following questions.

1 Which level of control does an elevated work platform come under?

2 List the four key requirements of solid construction.

1 ______________________________

2 ______________________________

3 ______________________________

4 ______________________________

3 Handrails for edge and void protection must be constructed at a height of:

a 0.9–1.1 m

b 2.0 m

c 1.0 m

d 1.2–1.5 m

4 Circle 'True' or 'False'.
It is not acceptable to use saw stools and/or step ladders to access a work platform. True False

5 Why should box gutters and valleys be walked on with great care?

______

______

______

6 A licence is required to install scaffold when the potential fall from the working platform is greater than:

a 2.0 m
b 1.0 m
c 2.4 m
d 4.0 m

7 List four points to consider when deciding to use trestle scaffold.

1 ______
2 ______
3 ______
4 ______

8 List two reasons why you should not use existing parapets, chimneys and brick/blockwork walls as fall protection or as tie-in for scaffolding.

1 ______
2 ______

9 The minimum width of a working platform where a potential fall is 2 m or greater is:

a 1.0 m
b 0.6 m
c 0.3 m
d 0.45 m

10 What is the importance of knowing the SWL of an EWP or scaffold work platform?

______

______

______

______

______

11 Circle 'True' or 'False'.

You need to work on a task that is 8 m off the ground and using an EWP. The only EWP available is a boom type that can reach 14 m in height. This is OK because you do not need a licence if working at heights less than 11 m.	True	False

12 What must be considered when installing hessian or shade cloth to scaffolding?

______

______

______

13 When a roof has a pitch greater than 35 degrees, what are the three options available to you for edge protection? How many of these must be in place?

1 ______

2 ______

3 ______

How many must be in place: ______

14 The code of practice covering excavation work requires that you put fall protection in place around any excavation work that offers a potential fall of more than:

a 0.9 m

b 1.0 m

c 1.5 m

d 2.0 m

15 What evidence might you need to be sure that a structure is capable of holding the required fall protection? Who would you ask for this information?

Evidence required: ______

Who to get it from: ______

WORKSHEET 3

To be completed by teachers	
Student competent	☐
Student not yet competent	☐

Student name: ______________________________

Enrolment year: ______________________________

Class code: ______________________________

Competency name/Number: ______________________________

CPCCCM2012: Work safely at heights – specifically, the element 'Access work area'

Also:

CPCCCM2010B: Work safely at heights – specifically, the element 'Access work area'

CPCCCM2010: Work safely on scaffolding higher than two metres – specifically, the element 'Access work area'

CPCPCM2055A: Work safely on roofs – specifically, the element 'Prepare for work'

Task: Answer the following questions.

1 Describe the difference between work positioning systems and travel restraint systems.

Work positioning: ______________________________

____

Travel restraint: ______________________________

____

2 List the three main types of fall-arrest systems.

1 ______________________________

2 ______________________________

3 ______________________________

3 The minimum distance a catch platform must extend beyond the work area is:

a 0.9 m
b 1.0 m
c 1.5 m
d 2.0 m

4 Who may install a safety net?

____

____

5 List the four main components of an IFAS.

1 ______________________________

2 ______________________________

3 ______________________________

4 ______________________________

6 How is the safe deployment distance of an IFAS calculated?

____

____

7 When doing up the screw on a karabiner, what must you do once it is tight?

____

____

8 An anchorage point designed for use with an IFAS by a single worker must be rated to:

a 6 kN

b 15 kN

c 1 kN

d 30 kN

9 Circle 'True' or 'False'.

An IFAS is considered inappropriate for most single-storey domestic housing construction because its safe deployment distance is greater than the likely fall heights. True False

10 You and your workmate have a combined weight of 150 kg. The tools and equipment you need weigh a total of 17 kg. The EWP available has a maximum SWL of 200 kg. What is the maximum weight of materials that you may take up with you?

a 23 kg

b 13 kg

c 33 kg

d 13 kg

11 When putting on (fitting) a harness, what are the two first things you should check?

____

WORKSHEET 4

To be completed by teachers	
Student competent	☐
Student not yet competent	☐

Student name: ____________________

Enrolment year: ____________________

Class code: ____________________

Competency name/Number: ____________________

CPCCCM2012: Work safely at heights – specifically, the element 'Conduct work tasks'

Also:

CPCCCM2010: Work safely on scaffolding higher than two metres – specifically, the element 'Conduct work tasks'

CPCPCM2055A: Work safely on roofs – specifically, the element 2 Prepare for work and 3 Install and use fall prevention system

Task: Answer the following questions.

1 After reading the section on ladders, list at least five reasons why ladders are not the preferred means for undertaking work at heights.

1 ____________________

2 ____________________

3 ____________________

4 ____________________

5 ____________________

2 When moving from one secure location to another, or transiting past an obstruction on a static line (such as a mounting post), how do you do so when using a double lanyard?

a Adopt a steady stance and remove both hooks at the same time then quickly reconnect both to the new point of security or line.

b Remove one hook, connect it the new point of security or line, then remove the other hook - one lanyard should be connected at all times.

c Twin or double lanyard systems are now prohibited in all states and territories of Australia except Queensland.

d Hold onto the rail or line and swap the one that is connected to the new point. The other lanyard should not be in use as it is a backup and should be hooked to some part of the front of your harness.

3 What is the purpose of a toe board on a working platform?

____

4 A working platform has been constructed for you to use. It complies with the appropriate Australian Standards but does not have toe boards and they cannot be fitted. What must you do?

a Refuse to carry out the work until a platform with toe boards is constructed.

b Do the work regardless, as the platform complies with the standards.

c Establish a 'no go' area beneath the platform for the duration of the work.

d 'Jerry rig' something of your own to replicate a toe board.

5 Circle 'True' or 'False'.

The lanyard on your IFAS is limiting your reach to the work you are doing from an EWP. You can undo it for the short period of time you need to reach that awkward spot, provided it is for not more than three minutes. True False

6 What must you consider before shifting materials, tools and yourself on to the platform extension of a scissor lift?

____

7 You are about to step off the edge protection platform and on to a sloping steel roof when you notice it is covered in dew. What should you do, and why?

What to do: ____

____

Why: ____

____

8 The only access to a working platform is by means of a correctly installed ladder. Tools and materials may be transferred up to the platform by:

a Carefully passing up long lengths, making sure no one is underneath

b Carrying them up the ladder, making sure to use only one hand to carry them

c Passing them up from platform to platform, making sure no one is underneath

d Either 'a' or 'c'

9 Describe the procedure for fixing battens to roof trusses.

____

____

____

____

____

10 Describe the procedure for installing safety mesh to a roof surface.

____

____

____

____

____

11 Describe the procedure for installing roof sheeting.

____

____

____

____

____

12 What is a 'recovery plan', and what are the seven points you should cover when creating one?

Recovery plan: ______________________________

____

Points to cover:

1 ______________________________

2 ______________________________

3 ______________________________

4 ______________________________

5 ______________________________

6 ______________________________

7 ______________________________

13 You have completed the job and must remove all waste and spare materials, as well as all tools and equipment. How will you bring these down to ground level?

a Passing down long lengths carefully, making sure no one is underneath

b Carrying them down a ladder, making sure to use only one hand to carry them

c Passing them from platform to platform, making sure no one is underneath

d Either 'a' or 'c'

14 When laying floor sheets to upper storeys, what potential fall risk remains despite following best practice steps, and how might you overcome it?

____

15 Circle 'True' or 'False'.

The primary aim of a recovery system for someone who has fallen and is suspended in their harness is to *get them down and lay them down as quickly and safely as possible.*	True	False

16 The correct angle to install a ladder is:

a 4 up and 1 out

b 3 up and 1 out

c 1 up and 4 out

d 5 up and 1 out

POWDER-ACTUATED TOOLS

2

This chapter covers the following topics from the competency 'Power actuated tools':

1 Background
2 Regulation of the use of powder-actuated tools
3 PA tools: the basics
4 Safe use of PA tools
5 Loading the PA tool
6 Firing the PA tool
7 Adaptors and accessories for PA tools
8 Removal of fasteners
9 Maintaining PA tools
10 Clean-up and stowage

Powder-actuated (PA) tools, also known as *explosive-powered tools* (EPTs), are designed to drive a fastener into or through material by means of a small explosive charge. In construction, this generally means the fixing of timber or light metal to steel or concrete. Although their use no longer requires a licence in any state or territory in Australia, they are recognised as a highly dangerous tool for which documented training is required. The purpose of this chapter is to assist both learners and instructors in achieving that end.

Overview

This chapter addresses the key elements of the following units:

- CPCCCM2007: Use explosive power tools
- CPCCCA3027: Set up, operate and maintain indirect action powder-actuated power tools

Prerequisites

- Plumbing, Gas fitting, Fire protection, Telecommunications and all Roofing courses: CPCPCM2043: Carry out WHS requirements
- CPCCWHS2001: Apply WHS requirements, policies and procedures in the construction industry

Background

Indirect powder-actuated tools are typically used in construction for tasks that involve fastening materials to various surfaces, such as concrete, steel or masonry. These tools use controlled explosive charges (powder-actuated loads) to drive fasteners, such as nails or pins, into hard materials. The 'indirect action' refers to the fact that the explosive force is transferred through a piston or other intermediary mechanism to drive the fastener.

When working with steel structures, indirect action powder-actuated tools can be used to fasten items such as steel studs, channels or brackets to steel beams, columns or other steel components.

In outdoor construction projects, these tools can be used to attach fence posts, deck framing, and other structural components to concrete slabs or footings.

Access plans, drawings and work instructions to determine the correct tools and equipment needed for the job. Using powder-actuated tools requires planning and communication with other personnel on-site.

By definition, powder-actuated (PA) tools are not firearms (see AS 1873.1 Powder-actuated (PA) hand-held fastening tools – Selection, operation and maintenance); however, they do rely upon the energy derived from a small brass cartridge not dissimilar to that designed for a .22 calibre rifle or pistol. In addition, they seem to increasingly resemble small automatic weapons. Furthermore, they do 'fire' a fastener, which, through either deliberate misuse or poor operator handling, can become a projectile capable of causing death or serious injury.

Due to this latter point, and the fact that PA tools are becoming ever more common on Australian work sites, it is critical that proper training be provided to operators. This is so even if you are only going to fire the odd fastener or two. Training should always be provided by skilled and experienced instructors using a range of tools on a variety of materials.

The purpose of this text is not to replace such instruction, but to supplement it: to provide a reference prior to, during and after undertaking training. It is also aimed at being a valuable resource for trainers. In providing detailed background material, examples and student self-evaluation tools, the text aims to cover and expand upon all the theoretical elements of the national competencies CPCCCA3027: Set up, operate and maintain indirect action powder-actuated power tools; and CPCCCM2007: Use explosive power tools.

As with any skill area, you will need to become familiar with a few terms and phrases relating to PA tools before informative instruction can begin. To this end, a short table of terms and definitions is provided in **Table 2.1**.

Regulation of the use of PA tools

Operating a powder-actuated (PA) tool is to be involved in high risk construction work. This is legislated in all Australian state and territory Work Health and Safety regulations (including Victoria's Occupational Health and Safety regulations). This means that before work can be carried out, a safe work method statement (SWMS) must be prepared (an example SWMS is provided in Appendix 1 at the end of this chapter). And although no state or territory requires a license to operate a **PA tools**, the following is also mandatory in all jurisdictions:

- before a person can operate a PA tool, their employer must provide training in the safe use of such tools generally and that tool specifically
- such training must be documented by the employer and countersigned by the operator.

This training, and the eventual use of PA tools, is regulated by the following Australian Standards:

- AS/NZS 1873.1 Powder-actuated (PA) hand-held fastening tools – Selection, operation and maintenance
- AS/NZS 1873.2 Powder-actuated (PA) hand-held fastening tools – Design and construction

TABLE 2.1 Key terms and definitions

Key Term	Definition
PA Tool	A direct or indirect acting tool that uses an explosive charge to drive a fastener into or through construction materials
Cartridge or Charge	A small brass .22 bullet like casing of explosive smokeless powder – the propellant
Fastener	A drive pin, nail, dowel, threaded stud driven by the action of the PA tool into or through material
Direct acting	A PA tool whereby the propellant's energy acts directly against the fastener
Indirect acting	A PA tool whereby the propellant's energy acts against an encapsulated piston which in turn drives the fastener
Base Material	The material into which the fastener will be fixed
Cycling	Action of bringing the piston and charge into place to ready the PA tool for the next shot
Misfire	A cartridge failing to fire due to faulty manufacture, cartridge misalignment
Authorised Person	The manufacturer, their nominee, or a qualified gunsmith that may repair a PA tool
Adaptor	A fitting designed to allow specific PA tools to be used in specialised situations.

- AS/NZS 1873.3 Powder-actuated (PA) hand-held fastening tools – Charges
- AS/NZS 1873.4 Powder-actuated (PA) hand-held fastening tools – Fasteners.

Note: At the time of publication, all the above standards have been withdrawn. However, as nothing has become available to replace them, it is advised that these standards should remain as your guide.

In addition, each state or territory WHS/OHS authority offer a set of regulations, guides, codes of practice, compliance codes and/or guidance material (see 'References and further reading' at the end of the chapter). It is important that you and your instructor ensure you are complying with any specific requirements applying in your state or territory.

KEY POINTS

- Operating a powder-actuated (PA) tool is high risk construction work and so a SWMS document must be prepared before any work is carried out.
- PA tools drive a fastener using a cartridge that looks like, but is more powerful than, those used in a .22 rifle.
- Users of PA tools must be trained on each specific tool.
- It is the responsibility of employers to provide training.

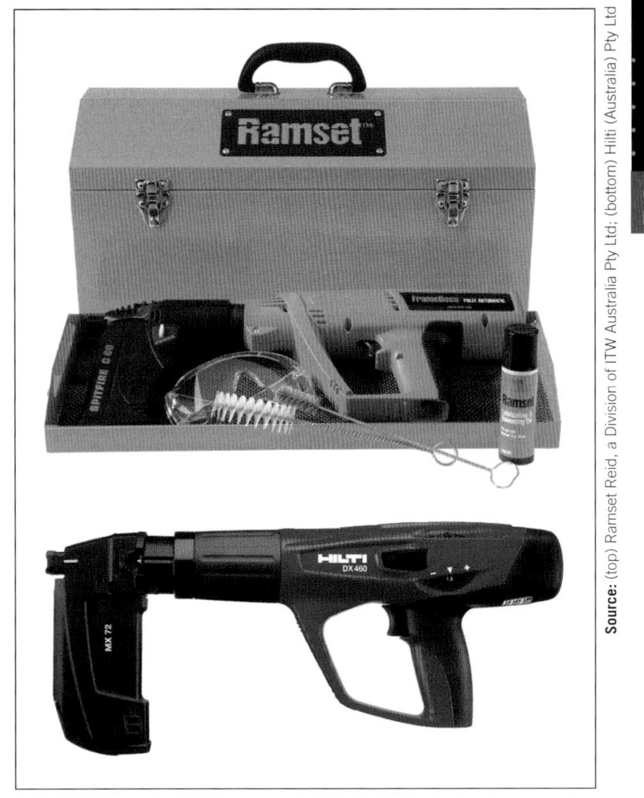

FIGURE 2.1 Modern fully automatic PA tools by the leading manufacturers Ramset and Hilti

Source: (top) Ramset Reid, a Division of ITW Australia Pty Ltd; (bottom) Hilti (Australia) Pty Ltd

PA tools: the basics

As stated earlier, all PA tools (Figure 2.1) use an explosive **charge** to drive a fastener against, into or through building materials. This explosive charge comes in the form of a small brass **cartridge** containing a carefully measured quantity of gunpowder, the number of grains or weight of powder determining the strength or force of the charge. It is from the explosion of this gunpowder that the tools derived the name **explosive-powered tools (EPT)**.

The term *powder-actuated* is currently favoured over *explosive-powered* due to the connotations of this latter phrase: 'explosive' sounding somewhat aggressive for what is, in effect, a very small charge. In addition, it suggests that the operator has training and experience in explosives of a demolition nature, which clearly is not the case.

The fastener itself is a drive pin, threaded stud, nail, dowel, rivet or similar object designed for use solely in a PA tool (see Figure 2.2). Depending upon requirements, **fasteners** may be driven into, or used to fix, materials in the following manner:

- timber to concrete or steel
- steel to concrete or steel.

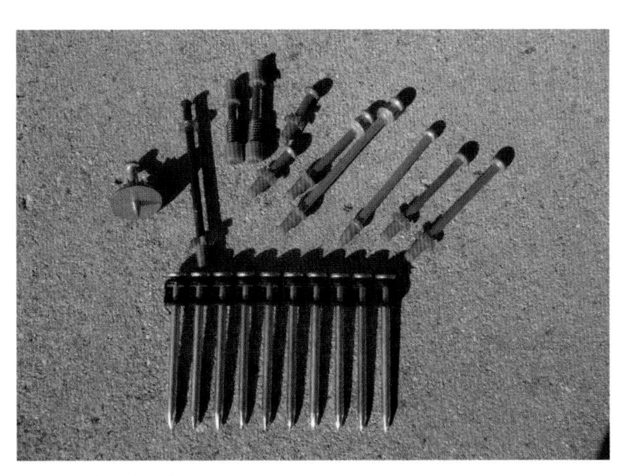

FIGURE 2.2 PA fasteners

Appropriate and inappropriate base materials

The material that the PA tool fastener is ultimately driven into is known as the 'base' material. Due to the expansive forces involved when a fastener is driven into the **base material**, anything that is brittle or excessively soft should not be used: brittle, because it may shatter, sending shards at the operator or other workers; and excessively soft, as the fastener may go clean through the material and free flight into an adjoining room or work area.

Appropriate base materials

Appropriate base materials include:

- mild steel
- concrete less than three years old
- brickwork* (except glazed bricks).

*Brickwork should generally be avoided as a base material, though it is not prohibited. Additionally, only certain solid bricks are suitable. Hollow bricks and hollow concrete blocks must not be shot into on any occasion.

Inappropriate base materials

PA tools are never to be used on the following:

- glass, tiles, marble, slate, natural stone, glazed brick, high-tensile steel, cast iron, welded joints, or areas of steel that have been torch cut (which can cause localised tempering), are brittle and may shatter and/or deflect the fastener. Hollow bricks and block work are also dangerous. This is because the fastener can pass through the blocks/bricks due to incorrect charge selection based on an initial test on a seemingly solid section
- the likes of plasterboard, timber and very thin steel should also not be used as a base material, but can be shot 'through' as the 'fixed' material (i.e. material that is being fixed to some suitable base).

Direct or indirect acting: an important difference

Only PA tools designed and manufactured in accordance with AS/NZS 1873 Powder-actuated (PA) hand-held fastening tools may be used in the workplace. Within these design rules there are two main types of tool: **direct acting** and **indirect acting**. *The importance of knowing the difference between direct-acting and indirect-acting tools, and being able to identify one tool from the other, cannot be overstated.* It can mean, and has meant, the difference between life and death.

Direct-acting PA tools (also known as 'high-velocity tools')

Direct-acting PA tools (Figure 2.3) release the energy of the cartridge directly on to the fastener in much the same manner as a rifle or pistol. As the energy is now effectively contained by the fastener, it passes down the **barrel** and exits the tool at a very high velocity – much as a bullet might (see Figure 2.4). These are therefore very dangerous tools, and their use is becoming limited on construction sites. They are also the tool involved in most PA tool-related deaths and injuries.

Source: Photo supplied by Jeff Millar

FIGURE 2.3 Direct-acting tools

Due to these high velocities, should the tool be mishandled, it is also possible that the fastener might ricochet off the fixing surface and strike you or a fellow worker in or around the work area. In an endeavour to prevent such incidents, it is required by regulation that an effective muzzle guard or protective shield be fitted to direct-acting PA tools according to the manufacturer's recommendations.

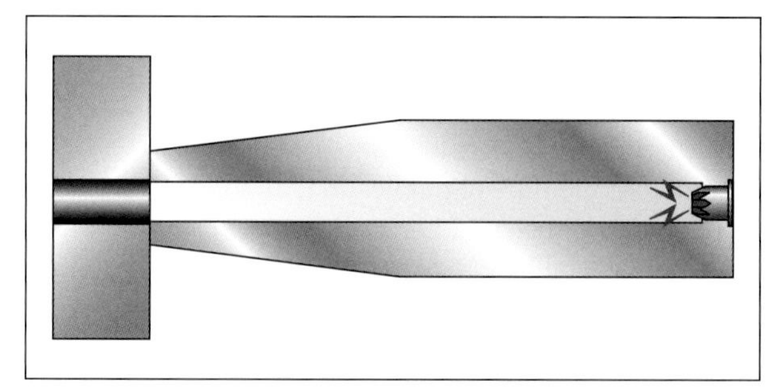

FIGURE 2.4 Energy transferred directly to fastener

ON-SITE

SAFETY NOTE

Research has shown that the ballistic characteristics (the damage done to people, animals or materials) of direct-acting tools more correctly replicates that of a 9 mm military pistol than the .22 cal. handgun that they are generally compared to given the size and shape the of cartridge (Frank et al., 2012).

Source: Shutterstock.com/Militarist

A further reason for the reduced presence of these tools on construction sites today is that they provide a single shot only. Each fastener, and then charge, must be placed individually by the operator prior to firing. This is time-consuming, and so the tools are fast becoming superseded by the **magazine**-fed, and safer, indirect-acting tools. In addition, neither of the two leading manufacturers and suppliers of PA tools in Australia currently produces a direct-acting tool.

On most commercial construction sites in Australia, direct-acting PA tools are prohibited by industry. The tools themselves, however, are not prohibited by law (nationally or by any state/territory) and so they are still to be found occasionally in operation on domestic sites.

ON-SITE

EXAMPLES

Fastener velocity for direct-acting PA tools can reach almost 600 m per second: enough to propel the fastener across a busy city street, through a window, and kill a person sitting at their desk. Alternatively, it can drive a fastener through a floor, then through some light ceiling material, and kill a person on the floor below. Unfortunately, both these scenarios have occurred.

Indirect-acting PA tools (also known as 'low-velocity tools')

Indirect-acting tools (Figure 2.5) differ from direct PA tools in having an encapsulated **piston** between the charge and the fastener. 'Encapsulated' means that the piston can only move forwards and backwards within the tool; it doesn't come out of the tool with the fastener. When the charge is activated, the resultant force is transferred to this piston, which in turn 'pushes' the fastener forward through the material (see Figure 2.6).

Source: (top) Ramset Reid, a Division of ITW Australia Pty Ltd; (bottom) Hilti (Australia) Pty Ltd

FIGURE 2.5 Indirect-acting tools

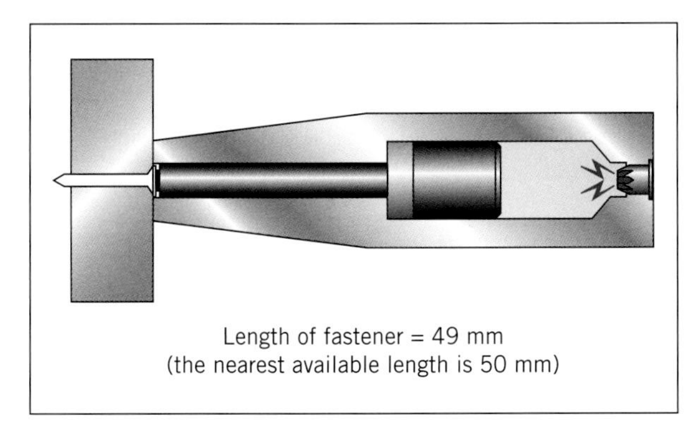

FIGURE 2.6 Force is transferred to the piston when the charge is activated

The result of this action is a fastener that exits the barrel much slower than it would from a direct-acting tool. Despite the reduced velocity, the fastener still enters the material with very similar energy to direct-acting tools while the piston is still pushing it. Once the piston loses contact with the fastener, there is very little energy in the fastener itself and so it cannot travel any great distance in free flight.

Key characteristics of indirect-acting PA tools include:

- the fastener travels at a relatively low velocity of 100 m per second or less
- due to the low velocity, most indirect-acting PA tools do not require a muzzle guard
- cartridges or charges are often supplied in strips, making the tool semi-automatic
- some recent tools can be fitted with magazines holding 10 or more fasteners, again making them more effective as a semi-automatic construction tool.

Tool labelling

Whether it is a direct- or indirect-acting tool, all PA tools must be labelled with the following information:

- manufacturer's name
- serial number
- model number
- a **misfire** warning, or the wording 'Refer to the operating instructions before using this tool'.

KEY POINTS

- There are 'appropriate' and 'inappropriate' base materials.
- Direct-acting PA tools are still legal in Australia, but are banned on some commercial building sites as they are much more dangerous than indirect-acting PA tools.
- All tools must be fully labelled to be legal on any Australian work site.

The main components of PA tools

There are a variety of PA tools available, and each manufacturer goes about producing their tools in slightly different ways; hence, the actual components of any given tool may vary. There are, however, some basic parts with which you should become familiar.

Direct-acting PA tools

These are simple tools consisting of very few parts. Some of the earlier tools comprised no more than a barrel and a screw-on breech for holding the charge, which was then struck by a hand-held hammer. The basic components of a direct-acting PA tool are shown in Figure 2.7.

Note: Direct-acting PA tools must not be fired unless fitted with the safety guard shown in Figure 2.7. This is because the likelihood of ricochets and flying debris from the face of the work piece is much higher with these tools than with the indirect-acting tools discussed next.

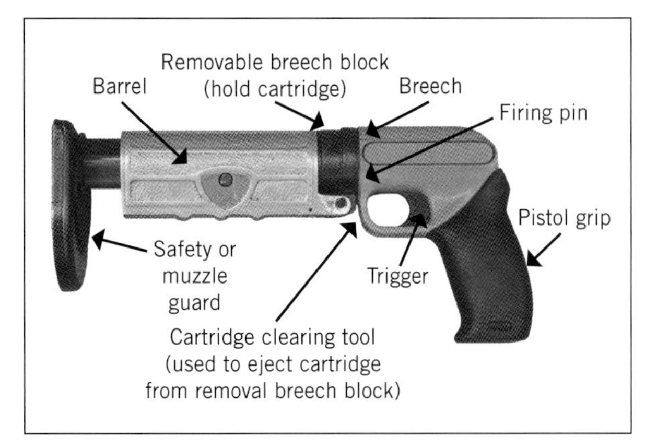

FIGURE 2.7 Direct-acting tool parts

LEARNING TASK 2.1

1 **Circle 'True' or 'False'.**
A direct-acting tool is the safest PA tool to use as there is no piston that can break and fly out of the barrel.
True False

2 **Which of the following base materials is appropriate for fixing into with a PA tool?**
a Mild steel
b Timber
c Cast iron
d Natural stone

Indirect-acting PA tools

In its earliest forms, this was still a fairly simple tool, having the addition of only a piston and the necessary components for holding it in place, yet allowing its removal for maintenance. Contemporary tools have tried to retain this simplicity (for ease of use, maintenance and general serviceability); however, they have added such things as silencers and magazine feeds (for both charges and fasteners). In addition, there are a raft of attachments available to aid the operator in aligning the tool or guiding specialist fasteners. Components of one of the more common units available are shown in Figure 2.8.

The charge: the heart of PA tools

The cartridge, or 'charge', of PA tools has been briefly described previously when the basics of the tools were discussed. Being the muscle behind the PA tool, it is important for operators to have a thorough understanding of the function and selection of charges. It is the inappropriate selection of charges that has, on occasion, led to the deaths and injuries inflicted by PA tools over the years.

PA tool charges are brass-cased cartridges specifically designed for use in PA tools. As stated earlier, in form they look not dissimilar to a cartridge for a .22 calibre rifle or pistol (see Figure 2.9). However, PA tool cartridges differ in two key respects:

1 They have no bullet or other projectile attached to them, and so are more like a 'blank'.
2 They are colour coded for differing strengths of charge.

Within each cartridge is a specific quantity of black gunpowder grains for the strength of shot for

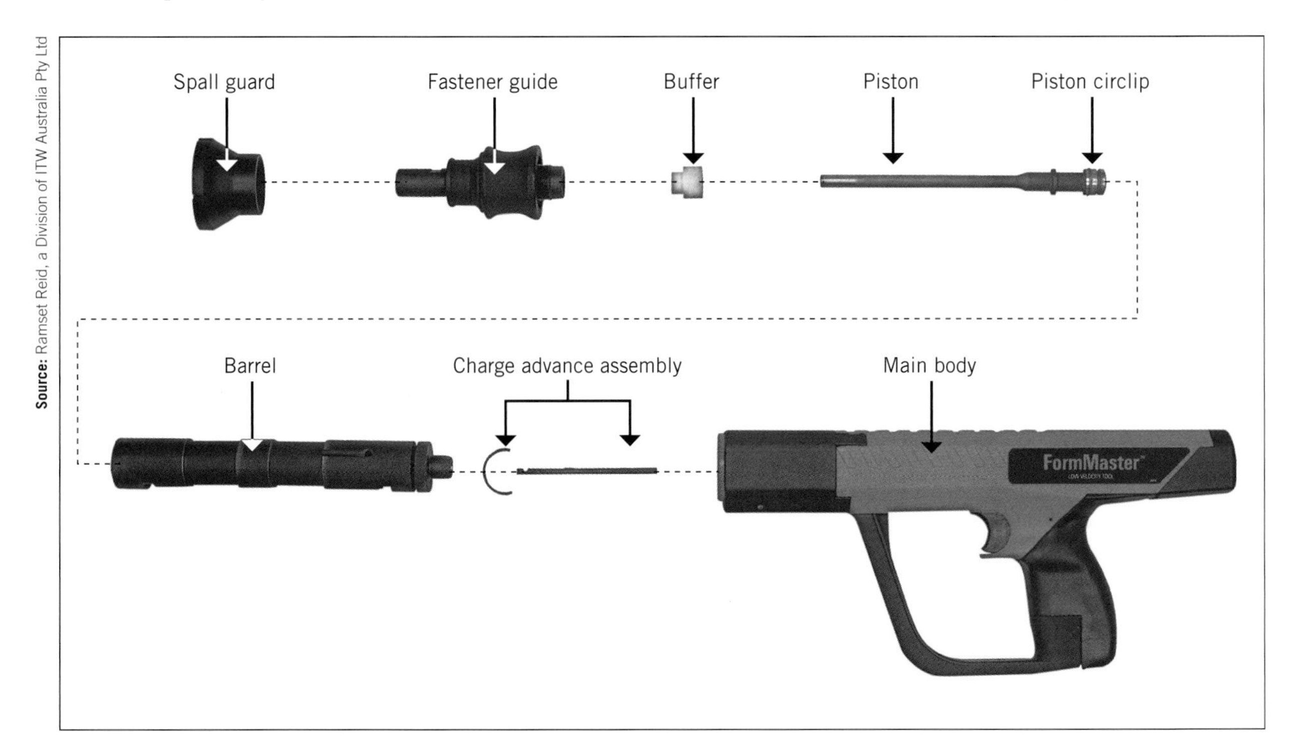

Source: Ramset Reid, a Division of ITW Australia Pty Ltd

FIGURE 2.8 Indirect-acting tool parts

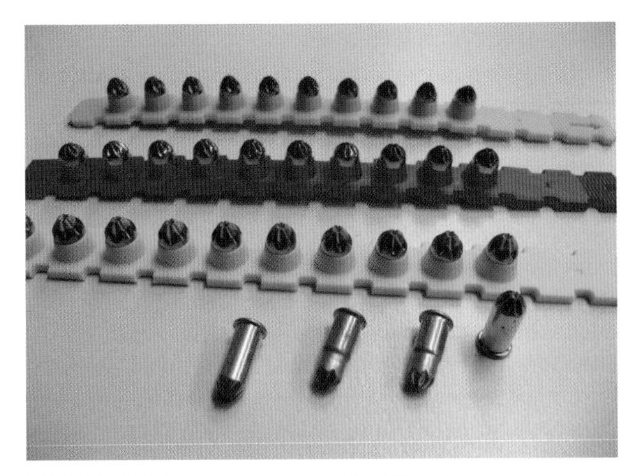

FIGURE 2.9 Colour-coded charges for PA tools

which it is colour coded. In addition, there is a small priming charge that is compacted into the rear of the cartridge around the rim (what is known as a 'rim fire' cartridge). When compacted further by the action of the **firing pin**, the primer ignites, setting off the main charge.

Should the primer fail to ignite when struck, the tool is said to have 'misfired'. In such cases, one of two things has occurred:

- the primer may have failed to ignite at all
- the primer may have partially ignited and be burning slowly.

In the latter case, we have a very dangerous situation, as the main charge may yet ignite and propel the fastener. How to deal with this situation as an operator is discussed under 'Misfire procedure' later in the chapter.

Depending upon the tool in use, PA charges are available in boxes of single cartridges, and in strips or discs for semi-automatic tools.

Strength of charges

As stated earlier, charges for PA tools are colour coded as a means of identifying their varying strengths (Figure 2.9). These colours are specified in AS/NZS 1873 Powder-actuated (PA) hand-held fastening tools, as set out in Table 2.2.

TABLE 2.2 Cartridge strengths and colours

Strength	Colour
Minimum	Grey
Very weak	Brown
Weak	Green
Medium	Yellow
Medium strong	Blue
Strong	Red
Very strong	Purple
Especially strong	White
Maximum	Black

In practice, only a few of these colours are readily available, or needed, for modern PA tools. They are listed below.

Weak	Green
Medium	Yellow
Strong	Red
Maximum	Black

The others are available, but often must be specially ordered.

For the novice, the difference between 'especially strong' and 'very weak' rightly seems a bit vague. The actual difference is a practical one, only measurable by application to a specific material set – that is, what you are fixing through and/or to. For this reason, the manufacturers require that the first test firing is always with the weakest charge available. This progressive process of test and elimination is dealt with more fully in the section titled 'Safe use of PA tools' later in the chapter.

ON-SITE

CAN YOU SEE COLOURS?

In the past, New South Wales required that operators of PA tools hold a licence, obtainable only after being trained and tested on their knowledge of tools, their application and the various charges available. Prior to acceptance for training, the prospective operator was tested for colour blindness. While the procedure is no longer mandatory, it is still relevant to the safe use of PA tools and should be conducted by employers and/or trainers on all trainees. Should the applicant/student fail the test, they should be advised not to attempt to use PA tools as they are exposing themselves, their fellow workers and the general public to unacceptable risks.

As a means of identifying charges should they become mixed, AS/NZS 1873 requires individual charges to be clearly marked with the following information:

- name and/or trademark of the manufacturer/supplier
- colour indicating the strength of the charge.

Storage of charges

While today's charges are manufactured to be quite safe when handled and stored correctly, they are still dangerous when deliberately mistreated. For this reason, live charges must be kept in a lockable container (Figure 2.10) marked with the words, '*Warning: Explosive Charges*'.

It is preferable that this container is held within the box for the PA tool itself. If this is the case, the outside of the PA toolbox must be likewise labelled with the above warning. Either way (in the PA toolbox, or

FIGURE 2.10 All storage boxes for charges must be marked with the words, 'Warning: Explosive Charges'

outside of it), the charges box should be locked when charges are not being accessed.

Only a trained operator should hold the key that gives access to live charges. In addition, in respect to live charges:

- never leave live charges lying around. Remove from the container only those needed for the immediate job. Return the unused ones to the lockable container
- never use a PA tool charge in a firearm
- never use a cartridge from a firearm in a PA tool
- never mix up charges from differing tools or with firearm cartridges*
- never deliberately strike a PA charge just to see if it will go off – assume that it will, with potentially deadly consequences
- never allow PA tool charges to become mixed with fasteners, tools or materials, or in your nail bag
- never leave live cartridges in a tool when transporting or carrying it from place to place.*

KEY POINTS

- The main component that is different between direct- and indirect-acting PA tools is the piston of the latter.
- The main chargers used on-site with modern tools are green (weak), yellow (medium), red (strong) and black (which is the maximum, available for some tools but infrequently used).
- Live charges must be stored in a lockable container.

* *Note:* Some early tools used charges of a smaller-diameter cartridge, some a longer cartridge, and others even a stepped or crimped cartridge (compared to the common cartridge of today's tools). Each must be used solely for the tool it is designed for. Failure to do so can lead to the cartridge itself rupturing upon firing, in so doing damaging or even fracturing the breech.

* *Note:* With the advent of magazine-fed, semi-automatic tools, it is permissible for the trained operator to carry a loaded tool short distances from firing point to firing point. However, cartridges should still be removed from the tool when moving from workplace to workplace (even within the one site), or from floor to floor. This issue is discussed further in the section titled 'Safe use of PA tools'.

Fasteners: types and uses

There are so many different types of fasteners available within the industry today it is outside the scope of this text to cover them all. There are, however, some general categories of fasteners, as well as the most commonly used ones. We will focus on these.

All fasteners designed to be used in PA tools are made from tempered (heat-treated) steel. In so being, they are very hard, yet (to some degree) brittle. This is due to the internal stresses that such steel contains. Fasteners must be hard to be able to withstand the high-impact forces generated when being driven against, and ultimately through or into, materials such as concrete or mild steel.

Fasteners are available in a variety of lengths, matching in the main the common lengths of nails and screws – that is, 40 mm, 50 mm, 65 mm, 75 mm, 100 mm and the like. Also note the following:

- Only fasteners designed to AS/NZS 1873 may be used in PA tools.
- For any given PA tool, use only fasteners designed for that specific tool. As with charges, some older tools have a different barrel diameter from contemporary tools and so require a particular-diameter fastener to match.
- Never 'adapt' a fastener to suit a tool for which it was not originally designed – for example, by swapping plastic guide tips or washers.
- Most fasteners fall into one of the following two categories:
  - drive pins
  - threaded studs.

  These two categories are discussed next.

Drive pins

Drive pins can look much like a normal hand-driven nail, or they may come with collars or washers, and even plastic tips. The plastic tips are provided by some manufacturers for guiding the pins down the barrel, while helping to clean it at the same time (Figure 2.11). It is also suggested that these tips reduce the likelihood of the fastener sliding upon impact with the receiving material, reducing the chance of ricochets. The collars or washers are used when fixing thin materials (such as metal sheet) to concrete or steel (Figure 2.11).

Drive pins are generally permanent fasteners, designed to give a non-removable fixing of one material to another. (Note, however, that manufacturers do produce a number of removable or temporary drive pins for formwork and the like.) The pins achieve this effect by means of **melding** the surface of the fastener with the surrounding material into which they have been driven with great force. This means that on a microscopic level, the concrete or steel immediately in contact with the drive pin has fused with the pin itself (see Figure 2.12). It is by this means, and not friction, that the fastener is held in place.

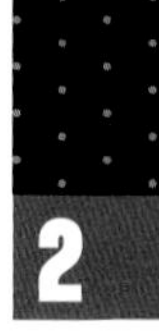

FIGURE 2.11 Drive pins with metal washer (left) and plastic tip (right)

Compression and melding of concrete around drive pin. Strength of hold is in part friction; in part from the concrete and steel bonding together.

FIGURE 2.12 Fusing of drive pin to concrete

To help achieve this melding of the fastener to the base material, drive pins are provided with either smooth shanks for concrete work, or 'knurled' shanks for fixing to thick mild steel (see Figure 2.13). Knurled shanks provide small ridges that effectively melt and weld themselves to the mild steel into which the fastener has penetrated. The smooth shanks work best in concrete, as it offers a broader surface to which the concrete can fuse.

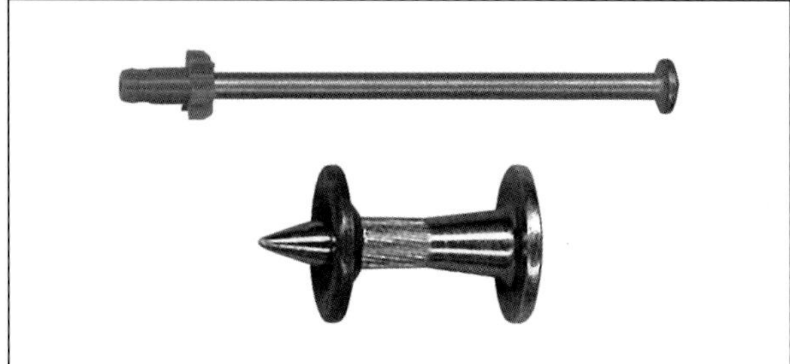

Source: Ramset Reid, a Division of ITW Australia Pty Ltd

FIGURE 2.13 Smooth (top) and knurled (bottom) shank drive pins

The manufacturer's stated length of a drive pin is its full length from head to tip.

Threaded studs

Threaded studs act like a permanently located bolt melded to the base material. They allow for the temporary or adjustable fixing of one material to another, much as welded studs would on a steel plate. Materials to be fixed then have holes drilled through them to align with the studs and are ultimately held in place by means of nuts and washers.

As with drive pins, threaded studs may come with knurled shanks for steel fixing, and smooth shanks for concrete. In addition, they may come with plastic tips, depending on the tool being used and the manufacturer.

Unlike drive pins, however, the length of a threaded stud is not as simple as its head-to-tip dimension. As shown in Figure 2.14 and Table 2.3, the information required when ordering threaded studs is more detailed. This information must include as a minimum the thread diameter and length, and the shank diameter and length.

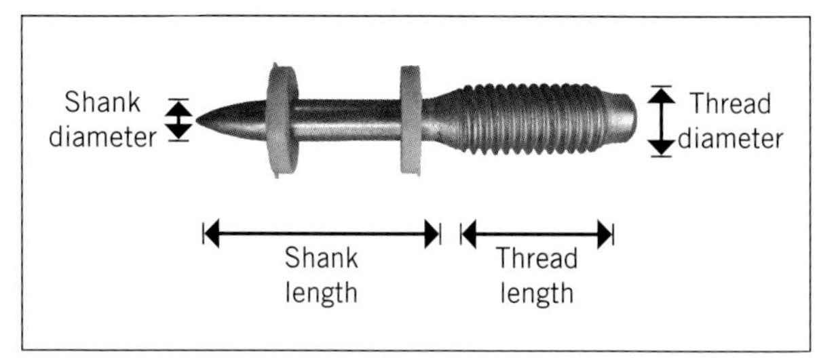

FIGURE 2.14 Threaded stud

TABLE 2.3 Threaded stud manufacturer's specification information

For use with	DX 36 M, DX 351, DX 460, DX E 72
Base materials	Concrete
Point type	Ballistic point
Washer size	8 mm
Washer type	Plastic
Thread diameter	M6
Thread length	20 mm
Shank length	27 mm
Shank diameter	4 mm
Weight	6 g
Material composition	Carbon steel shank HRC 55.5 ± 1
Corrosion protection	Galvanised zinc coated 5–13 μm

Determining the correct length of fasteners

Having decided either to fix or bolt your material permanently, and having chosen your fastener type accordingly, the next step is to determine its correct length. A number of factors must be taken into account when making this judgement:

- the type of base material into which you are fixing
- if steel, the thickness of this base material
- the thickness of the material to be fastened
- the available lengths of fasteners.

The following text outlines the calculations for determining the length of drive pins into steel and concrete. The manufacturer's stated length for drive pins or nails is from the top of the head to the tip of the point (not counting any plastic tips), as shown in Figure 2.15. The calculated length is therefore the actual required length of the pin.

With threaded studs the approach is slightly different, as the information required is more detailed (see Figure 2.14 and Table 2.3). This will be explained more fully at the end of the section.

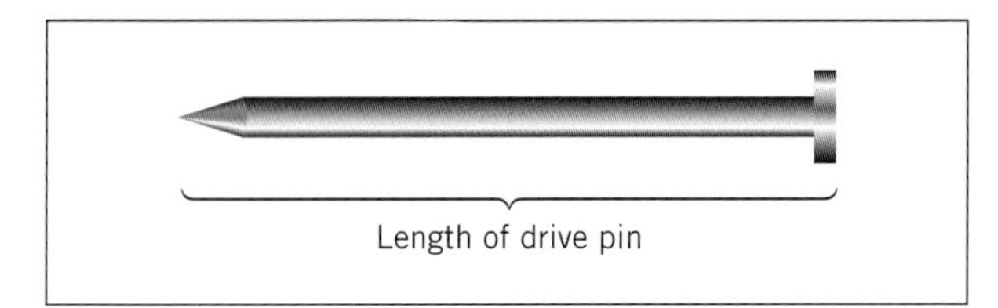

FIGURE 2.15 Drive pin measurement points

HOW TO

DETERMINING CORRECT FASTENER LENGTH

The calculations for determining the right length of fastener for each type of base material are set out below.

Steel

In most cases with steel as the base material, the fastener will penetrate through the metal. In this scenario, the length of the fastener is the sum of the following (see Figure 2.16):

- the thickness of the material being fastened, plus
- the thickness of the base steel into which the fastener will be driven, plus
- an allowance of 6 mm or more for the point to protrude through the base steel.

Length of fastener = 35 + 8 + 6

Length of fastener = 49 mm (the nearest available length is 50 mm).

Fixing into especially thick or solid steel, and/or very thin steel, requires a slightly different approach.

Thick steel

On occasion, the steel base material will be too thick for a fastener to be driven through it. In such cases, allow for a minimum penetration into the base metal of 12 mm. In effect, you are driving into solid steel, and on these occasions you should select a fastener with a knurled shank. This will improve the effective melding of the surface of the fastener with the steel surrounding it.

Thin steel

Fixing into thin steel as your base material is dangerous for two particular reasons:

1. The fastener can easily be over-driven and so fly free from the back or underside of the material. This can lead to direct or ricochet injuries to yourself or others.
2. The light sheet steel can deform under the impact of the fastener. This can cause the fastener to ricochet or deflect off, rather than penetrate the material. Again, this can lead to injuries to yourself or others.

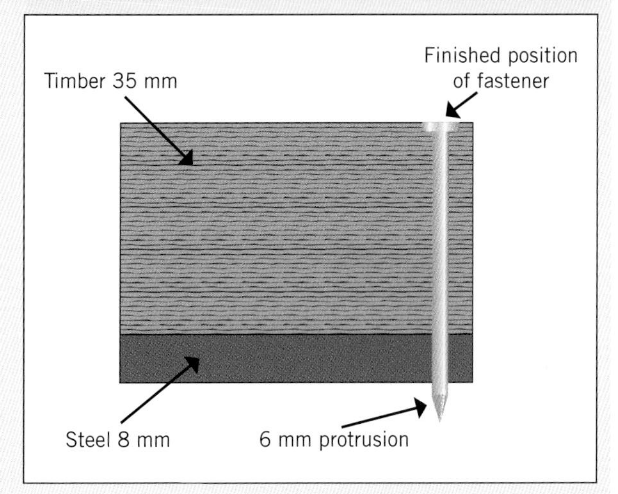

FIGURE 2.16 How to determine the required length of a drive pin

Special fasteners have been designed for steel down to 3 mm thick. Only these particular fasteners should be used in such cases.

Never attempt to fasten to steel 4 mm thick or less unless using fasteners specifically designed for this purpose.

Concrete

When fixing into concrete, the concrete must always be significantly thicker than the penetration depth of the fastener. As a rule, the thickness of concrete should be:

> 100 mm, or three times the shank penetration depth of the fastener, whichever is the greater.

If, for some reason, this is not the case (such as when fixing large-diameter fasteners to a thin precast wall or suspended slab, for example), *PA tools should not be used.* In such cases, the back or underside of the concrete can 'blow', sending both shards of concrete and the fastener itself into the adjoining room or space. Needless to say, this is less than desirable!

>>

>>

Having satisfied yourself that the concrete is of an appropriate thickness, the calculation for the length of the fastener is as follows:

- the thickness of the material to be fastened, plus
- six to eight times the shank diameter of the fastener.

As a formula, this is expressed as:

$$\text{Fastener length} = T + (6 \sim 8 \times \text{shank } ø)$$

Where:

T = thickness of the material to be fastened

ø = diameter of shank

As most threaded studs and drive pins have a shank diameter of 3.8 mm, the 'rule of thumb' is to allow for 25–30 mm penetration into the concrete (see **Figure 2.17**).

For example:

$$\begin{aligned} \text{Fastener length} &= T + (6 \sim 8 \times \text{shank } ø) \\ &= 45 + (6 \sim 8 \times 3.8) \\ &= 45 + 22.8 \sim 30.4 \\ &= 67.8 \text{ mm} \sim 75.4 \text{ mm fastener} \end{aligned}$$

Our 'rule of thumb' suggests a penetration of 25–30 mm, so of the available fasteners, 75 mm is the appropriate length.

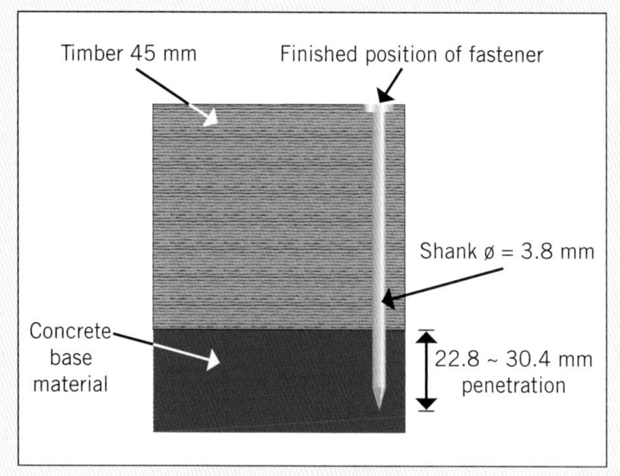

FIGURE 2.17 How to determine the required length of threaded stud or drive pin into concrete

DETERMINING THE CORRECT LENGTH OF THREADED STUDS

The approach with regard to penetration is the same as shown for drive pins, whether this is into concrete or steel. However, the information you will need to derive is more detailed, including the shank and thread lengths as separate dimensions. The thread length must include an allowance for the nut and washer. **Figure 2.18** uses the same material information as shown in **Figure 2.17**. Note that what is required now is a short shank length that will penetrate the steel, and a thread length long enough to allow the timber to be bolted into place. A similar approach is required for fixing to concrete.

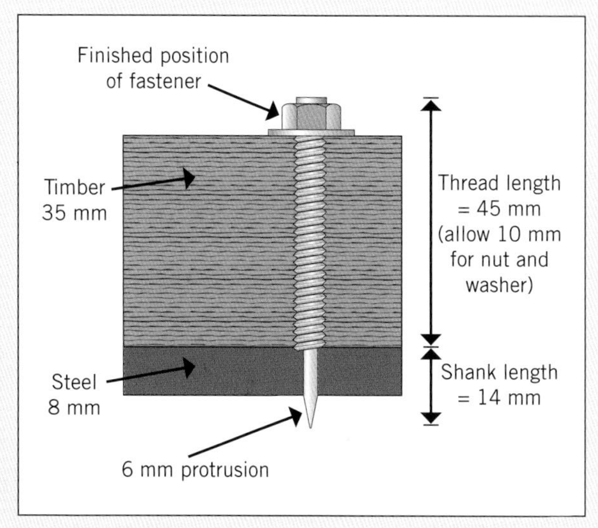

FIGURE 2.18 How to determine the required length of threaded stud

LEARNING TASK 2.2

1 **Circle 'True' or 'False'.**
You should never use a PA tool to fasten to steel less than 3 mm thick.
True False

2 **When fixing into concrete, the concrete must be the greater of:**
 a 75 mm thick or 3 times the shank penetration depth
 b 100 mm thick or 3 times the shank penetration depth
 c 75 mm thick or 4 times the shank penetration depth
 d 100 mm thick or 4 times the shank penetration depth

Supplementary activities:

- Go to https://enchroma.com.au/pages/colour-blind-test and test yourself for colour blindness.
- Under supervision (teacher or competent tradesperson/employer), strip and reassemble a PA tool.

COMPLETE WORKSHEET 1

KEY POINTS

- Fasteners are hard yet brittle, and when snapped they tend to fly off at very high speeds.
- Only use fasteners designed for use in that tool – never adapt them.
- Calculate the length of your fasteners with care – too long or too short will give equally poor fixing.

Safe use of PA tools

There are a number of steps that you must take prior to even loading a PA tool. Each step must be followed fully and carefully, as not doing so can lead to death or serious injury of yourself, fellow workers or others in the surrounding area.

ON-SITE

SAFETY NOTE

Currently there are no codes of practice nor any guidance material available from any state or territory authority regarding the safe operation of PA tools. Aside from the information contained in this chapter, you should be guided by the user's manual provided by the manufacturer.

Step 1: establish a 'no go' area

Before firing can begin, you must first establish an area immediately around, and at times beneath, your work zone that is clearly defined as a **'no go' area**. A 'no go' area is a clearly defined area into which people, other than those required to complete the task(s) involved, are restricted from entering. In this case, they are the PA tool operator and the **spotter**.

Wherever possible, the 'no go' should encompass a 6 m radius clear area around the firing position. The establishment of this area must include signage, and the addition of a spotter is highly advisable. A spotter is someone who acts as your eyes and ears, looking around to make sure no one has entered the 'no go' area, or is otherwise in a dangerous position prior to you firing the tool. Note that you should never use a PA tool while on your own.

Signs must be of the dimensions and colours shown in Figure 2.19.

In addition, you may choose to use hoarding, orange barrier mesh, bright orange witches hats (safety or marker cones) and other means of clearly demarking the work area.

On multi-level construction, domestic or commercial, you must install this same signage in the area below where you are operating the tool. Additionally, you must install a spotter on the outside of this area to ensure that other workers or passers-by do not inadvertently enter the area without you being aware. This same approach must be adhered to if working on a rooftop.

Before firing, your spotter(s) or assistant must call out 'All clear' to let you know that the zones are safe.

Step 2: check the work area, task and set-up

Before even bringing the equipment to the firing location, you must check the following:

- there is no evidence of explosive or flammable gases, dusts or vapours
- it is not a zone of compressed air
- it is not a zone of extreme heat.

FIGURE 2.19 PA warning sign

Having satisfied yourself of the above, you may bring the equipment to the work area. While it is perfectly acceptable for an assistant to help bring equipment to the firing location, only the trained operator should open the box and handle the tool itself.

Checking materials for suitability

You must now ascertain that the materials you are required to fix, and the base material you are required to fix to, are suitable. While you will find a list of appropriate and inappropriate base materials earlier in the chapter, it is sometimes necessary to conduct a suitability test. This simple test should always be conducted when you have not used a particular material before. The procedure is as follows:

1. Wear eye protection.
2. Take a fastener without a plastic cap (or remove the cap) and hold it on the material to be tested as you would a nail to be driven into wood.
3. Strike the fastener head firmly with the hammer (again, as if nailing a piece of wood).
4. Check the point of the fastener: if damaged (bent or rounded), then the material is too hard.
5. Check the surface of the base material: if the fastener has left little or no impression, then the material is too hard; if it has sunk in several millimetres, then it is too soft; if it has cracked or chipped, then it is too brittle.

The material is suitable when the fastener is undamaged, and the material shows a clear but small indentation from the point of the fastener (about 0.5 mm deep).

With respect to concrete that is over three years of age, or that has been pre-stressed or post-tensioned, you should make enquiries as to its suitability with the site engineer and/or the manufacturer/supplier.

Checking behind the base material: pipes, wiring and voids

An extension of the suitability check is to ensure that there are no pipes or wiring hidden behind the material into which you intend to fire. Ways of determining if such risks are present include:

- look for service indicator signs for gas pipes or electrical lines (Figure 2.20)
- check with the site engineer, foreperson and relevant tradespeople working on the site
- check service diagrams for the construction you are working on.

In addition, you must check that the material you are firing into is thick enough not to allow an over-powered shot to penetrate right through and into another space. This could be another room, a lower floor or a street, which might be occupied by other workers or the general public (see 'Special considerations' later in the chapter). Where this is the case, or there is even a remote possibility of such

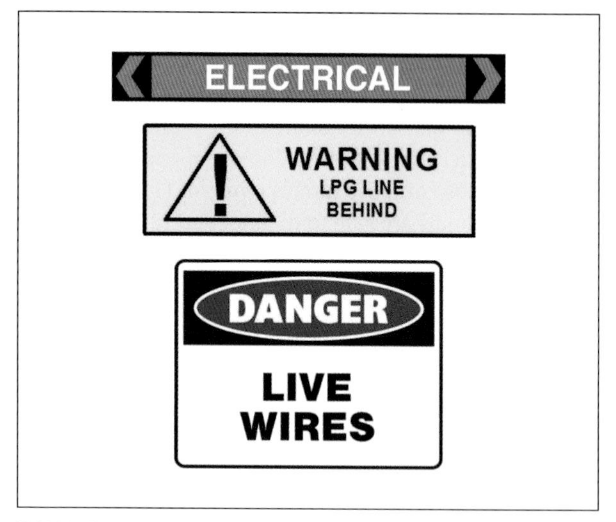

FIGURE 2.20 Service indicator signs for gas pipes or electrical lines

occurring, you should discuss with the site supervisor possible options, including:

- establishing 'no go' areas behind the material or walls being worked on
- ensuring that areas below or behind materials are closed to the general public
- undertaking PA work at times when other workers/the general public will not be present
- installing containment shields behind the base material/wall/floor
- using an alternative fastening system.

Step 3: check your tools and personal protective equipment

Before loading or firing a PA tool, you must first check it for serviceability. This involves following the manufacturer's instructions to complete a basic stripping of the tool in order to conduct the following checks:

- the tool is not loaded with either a charge or fastener
- the barrel is clean and clear of obstructions
- the tool is lightly lubricated where required
- if fitted, the position, condition and connection of muzzle guards or **adaptors** are as required
- the cocking and firing action is working correctly (without loading a charge or fastener).

Now check your personal protective equipment (PPE). PA tool operators and their assistants must be wearing the following as a minimum:

- safety glasses (to AS/NZS 1337.1 Personal eye protection – Eye and face protectors for occupational applications)
- hearing protection (to AS/NZS 1270 Acoustics – Hearing protectors).

In addition, it is advisable to wear hard hats, work boots with full leather uppers, and high-visibility vests.

Step 4: loading and test firing

Having created a safe 'no go' area, and made preliminary checks to ascertain the suitability of the materials, and the serviceability of the tool, you should now do a series of test firings.

The purpose of these tests is twofold: (1) it confirms that the materials are safe to fix into; and (2) it determines the strength of charge required for that particular range of materials and particular fastener. Usually, no more than three test firings are required (one under, one over, one on).

KEY POINTS

- Where possible, establish a 6 m radius 'no go' area and use a spotter.
- Check work zone and surface for suitability and/or dangers – that is, gases, hollow or thin base material, pipes and electrical fittings and inappropriate base materials.
- Check tool serviceability.
- Conduct a test firing.

Loading the PA tool

Given the raft of PA tools available today, this is very much a case of: 'Follow the manufacturer's instructions'. Some generalised basics for loading the PA tool (Figures 2.21–2.24) are as follows:

- Never point a PA tool, loaded or not, at any part of yourself (don't stare down the barrel) or anyone else.

FIGURE 2.21 Placing fastener into direct-acting tool

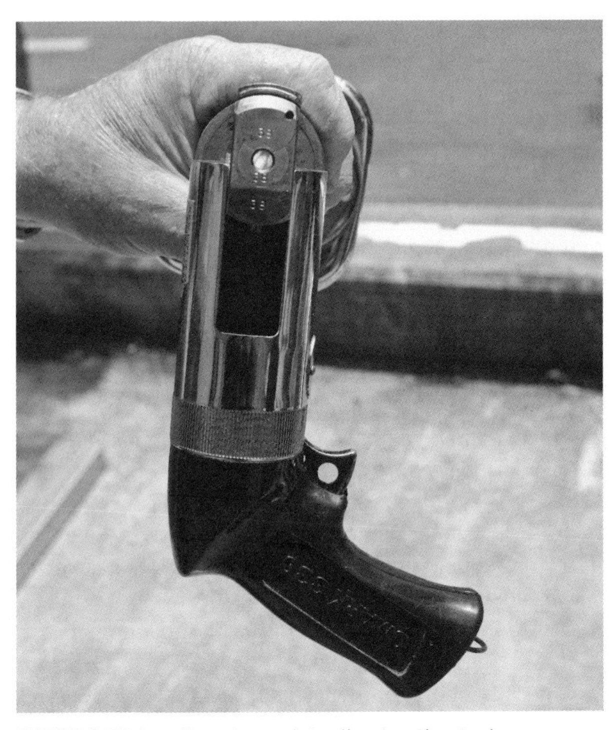

FIGURE 2.22 Loading charge into direct-acting tool

FIGURE 2.23 Placing fastener into indirect-acting tool

FIGURE 2.24 Loading strip charge into semi-automatic indirect-acting tool

Source: Ramset Reid, a Division of ITW Australia Pty Ltd

- Determine the fastener type and size, as per previous sections.
- On manual tools, load the fastener before the charge (or **cycling** the next charge into place on semi-automatic tools – see 'Cycling the tool' later in the chapter).
- Only load the PA tool in the immediate area of the place you are going to use it.
- Only load the tool when you are about to use it; do not load and then leave it.

As outlined earlier, always use the weakest charge available first when test firing, then progress upwards. Modern tools often have power adjustments on them, which may be effective if the second shot is slightly over-powered (see 'Power adjustment', below).

Power adjustment

Many modern PA tools have the capacity to reduce the force delivered by a given charge. This is done by changing the position of the piston (and hence the expansion chamber in the end of the piston) with relation to the charge itself. Normally the charge will sit tight in the chamber and so deliver its full charge to the piston. The adjustment mechanism available on modern tools pushes the piston away from the charge by means of a threaded rod and adjustment screw (Figure 2.25). This effectively enlarges the expansion chamber and so reduces the force applied to the piston.

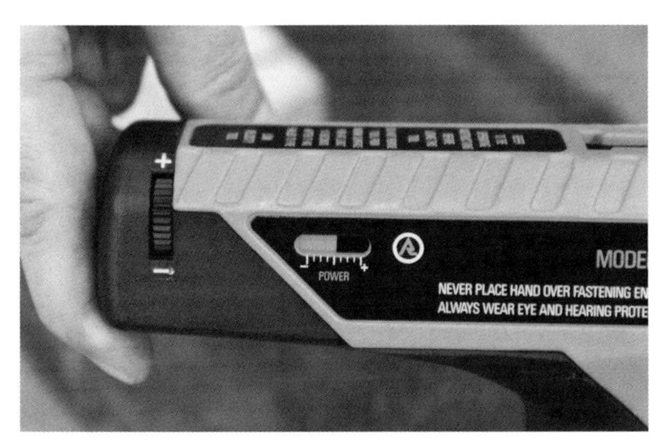

FIGURE 2.25 Power adjustment screw

With direct-acting tools, this same effect may be achieved by simply pushing or 'rodding' a fastener down the barrel. Once again, the area for the charge to expand into is increased and so the force received by the fastener is decreased.

Understanding and applying the above principles are important when fixings will remain visible to clients at the conclusion of the job. It allows you to 'tweak' the charge so that fasteners will finish flush, slightly proud (raised from), or appropriately below, the material surface as required.

Cycling the tool

Most single-shot, semi-automatic PA tools require that you 'cycle' the tool prior to firing. Modern fully automatic tools cycle automatically after each firing. Cycling brings the piston and charge into place and readies the trigger and firing pin for the next shot. This is also sometimes referred to as 'cocking' the tool. Without cycling, the tool will not fire. The action of cycling differs from tool to tool, and manufacturer to manufacturer, but generally it means sliding a section of the barrel, or its surrounds, forwards and then back into place.

With a tool that needs cycling prior to firing, the following procedure is considered best practice as it means you are not inserting fasteners into a 'loaded' barrel:

1. Insert the fastener into the barrel.
2. Cycle the tool to bring the charge into place.
3. Locate and fire the tool (following the procedure outlined below).
4. Insert the next fastener into the barrel.
5. Cycle, locate, fire.
6. Insert, cycle, locate, fire, etc.

Note: Only cycle a PA tool at the location of the next firing. *NEVER walk around with a tool that you have cycled*, as you are now effectively carrying a loaded tool.

Fully automatic tools are just that – automatic; that is, they automatically bring another charge (and on some tools, a fastener also) into place ready for firing at the next pull of the trigger. Cycling is therefore not necessary.

KEY POINTS

- Only load when you are ready and in position to use the tool.
- Never point a loaded tool, or one that you are in the process of loading, at any part of yourself or towards others.

Firing the PA tool

The PA tool is now ready to fire. The procedure for firing the tool is set out in the 'How to' box on page 78.

ON-SITE

SAFETY NOTE

Modern fully automatic PA tools (loading both fastener and cartridge, and cycling the tool) are specifically designed to disallow 'bump-firing' (also known as contact trip actuation, bounce-fire or simultaneous discharge). Bump-firing is when you hold the trigger and 'bump' the surface to fire the next fastener. Bump-firing is a dangerous practice common with pneumatic nail guns, and is an action made deliberately unachievable by the manufacturers. To attempt to alter or modify an automatic PA tool so as to achieve bump-firing is a prohibited act punishable by both state and federal laws.

HOW TO

FIRING THE PA TOOL

First check that you are wearing the appropriate hearing and eye protection (see 'Step 3: check your tools and personal protective equipment' on page 75). Then, in bringing the tool to the surface of the work piece, do the following:

1 Clear the surface of the materials you are fixing and/or fixing into of any dust or debris.
2 Check again that there are no wires or pipes you might be inadvertently driving into behind the work piece.
3 Fire no closer than 13 mm from the edge of steel, or 75 mm from the edge of concrete or brickwork (see Table 2.4).
4 Make sure any subsequent fasteners will be not less than 25 mm apart for steel, or 150 mm* apart for concrete (see Table 2.4).

TABLE 2.4 Fastener spacing and edge distances

Material	Spacing (mm)	Distance from edge (mm)
Steel	25	13
Concrete	150	75

5 If, for some reason, you have to remove, change or otherwise adjust a muzzle guard or barrel extension, you must remove any charge(s) first.
6 *Direct-acting tools*: ensure the muzzle guard is correctly positioned on the barrel. Unless it is rotated specifically to allow for getting closer to the side of the channel or beam, the guard barrel must be central to the guard.
After completing steps 1–6:
7 Hold the tool with both hands and place it perpendicular to the surface of the work piece (Figure 2.26). *Never fire a PA tool at an angle to the surface.*
8 Bracing your feet as shown in Figure 2.26, push the tool against the work face until the safety trip on the barrel has been pushed in.

Source: Ramset Reid, a Division of ITW Australia Pty Ltd

FIGURE 2.26 Hold the tool with both hands and place perpendicular to the surface – your feet should be braced

9 Ensure your arms are slightly bent and not tense; this will reduce the impact on your body of the recoil (particularly from older tools). *Never overreach with full arm extension and your elbows locked.*
10 Look or call to your spotter/assistant for confirmation that the area is clear for firing.
11 Your spotter/assistant must call back, 'All clear', in a loud voice. (If using a spotter below your work area, he or she must echo this call before you fire.)
12 While maintaining pressure against the work piece, call out loudly, 'Firing', and then squeeze the trigger.
13 Assuming the tool fires correctly, lift it clear of the work piece. Continue to point the tool away from you and/or your assistant (or anyone else) and check the seating of the fastener. From this, you may determine whether you need a stronger charge or not.

Repeat these actions for all subsequent shots.

Warning: If this first shot has not seated the fastener to the correct depth, *never place the tool over the fastener and 're-shoot' it with another charge.* This can cause severe damage to the tool (the piston can bend or break), and in the worst case, may shatter the barrel itself and kill or severely injure the operator. Neither of these is a desirable outcome.

* *Note:* The standard and some codes of practice state that fixings in concrete may be as close as 75 mm apart, though 150 mm is the manufacturer's stated distance.

Misfire procedure

In rare cases, the cartridge may fail to fire due to a fault in manufacture, or, in the case of semi-automatic tools using collated cartridges (cartridges joined together in strips), the cartridge may have been misaligned. This is known as a misfire.* Generally, this is because the primer has failed to ignite when compacted by the firing pin. However, as stated previously when discussing charges in general, you may be faced with a 'slow-burning' primer – that is, the charge may yet explode unexpectedly. The misfire procedure is outlined in the 'How to' box on page 79.

Jammed tools and live charges

Very rarely, a PA tool can jam with a live charge still in position. In such cases, once you have recognised the live jam, you must make no further attempts to adjust

* *Note:* Technically, misalignment of the cartridge is not a misfire; however, as the operator, you won't know this until you have removed or realigned the strip. You must therefore carry out the misfire procedure regardless of what you 'think' it might be.

HOW TO

PROCEDURE IN THE EVENT OF A MISFIRE

It is critical for the safety of yourself and your spotter/assistant that you always take the following action if you experience a misfire:

1. If you have pulled the trigger and the charge has not exploded (i.e. it has misfired), *do not let off the pressure*. Continue to hold the tool firmly against the work piece.
2. Count to 10 slowly.
3. Assuming the charge has not gone off during the 10-second count, remove the tool while being especially careful not to point it in the direction of yourself or others.

For single-shot tools:

4. Reposition the tool and attempt to fire the same cartridge again.
5. If it fails to fire again, maintain pressure and count to 10 once more.
6. Then open the breech and remove the charge. In so doing, do not place yourself in line with the charge; rather, extract it and hold it so you are looking at the side of the charge, rather than the front or back. In this way, should the charge go off unexpectedly (it is highly unlikely at this point), it is inclined to fly to your left or right, rather than back at your body or face.
7. Place the charge in a bucket or cup of water.
8. Retrieve the cartridge later and return it to the supplier for appropriate disposal. *Do not* put them in the bin or leave them lying around the workplace. *Do not* mix them up with live cartridges.

For semi-automatic and automatic tools:

4. Check the positioning of the strip in the tool.
5. If misaligned, align correctly and go through the firing procedure again.
6. If not misaligned, and it was a clear misfire, remove the strip and put it in a safe place isolated from other cartridges. Return the whole strip to the manufacturer for correct disposal.*
7. If the cartridge or cartridge strip cannot be removed normally (i.e. without undue force) then you must contact the manufacturer for the tool to be collected by them and repaired. Under no circumstances may you attempt to carry out this repair yourself.

* *Note:* Currently, no manufacturer requires you to place the whole strip in water. However, nor are you allowed to attempt to remove the offending cartridge from the strip. The whole strip must be disposed of as stated above.

the tool or remove the charge by force. You now have a very dangerous tool that can *only* be serviced by an **authorised person** (i.e. the manufacturer, their nominee or a qualified gunsmith).

It is best practice at such times to call the supplier or manufacturer and have their representative come and collect the tool from the workplace. In the meantime, you should lock the tool away in a location where, should it discharge inadvertently, it can do no harm.

If you must transport the tool yourself (which may be the case in outlying rural areas – in urban areas, if you do so, either you or the supplier are not taking the situation seriously enough), only do so in a locked storage container. In addition, you should locate the container in the vehicle such that should the tool discharge in transit, it will do no harm. (That is, ensure it is not pointing at the driver, cabin, passengers or following vehicles, which means pointing it across the vehicle or at the firewall in front of the passenger.)

Special considerations

There are a number of things that must be considered when firing a PA tool. It is critical that you are knowledgeable about all these issues before being considered a trained operator. Nothing may be taken for granted with PA tools: each issue outlined below has led, when not followed or taken into account, to incidents causing injury or at least a 'near miss'.

Firing into pre-drilled holes

On occasion, you may need to fire through pre-drilled holes to fix particularly hard or thick materials. This is acceptable only in the following circumstances:

- an approved adaptor or guide is fitted, which is specifically designed for that purpose and for that specific tool *(NEVER attempt to adapt such a fitting to a tool it was not designed for)*
- the point of the fastener is located in the pre-drilled hole prior to firing.

Note: NEVER attempt to fire a PA tool by eye or best guess into a pre-drilled hole, disc or washer. Doing so can lead to dangerous ricochets, deformation or even shattering of the barrel and/or internal piston. It is a very dangerous practice.

Firing into brickwork (solid bricks only)

Fixing into brickwork with a PA tool is not recommended, as generally the bricks are too small for an appropriate edge distance to be maintained. In addition, only solid bricks can be fired into and these may be hard to identify. Further, many solid bricks are too brittle to safely fire into without them shattering. Careful testing should be done prior to attempting to fix into this form of base material.

If you do determine that the material is suitable, then note that brick walls include mortar joints, which are soft – much softer than the bricks or blocks themselves – and not always are the joints fully pointed (filled with mortar front to back). This means that if you accidentally fire a PA fastener into the joint, it can pass clear through. Single-skin walls (walls only one brick thick) are particularly dangerous, as the fastener can fly into the adjoining room or space. In addition, if you partially hit a joint, the fastener can deflect and ricochet. So, while not prohibited, brickwork should be avoided as a base material.

When fixing to brick or blockwork:

- never fire into mortar joints – only into the blocks or bricks themselves
- when walls are rendered or painted, identify joints by measuring and then testing with a hand-driven nail
- unpainted rendered walls can be sprayed with water. This will generally show up the joints as darker lines.

Fixing into grooves and channels

There will be times when you will need to fix into or through 'C' or 'H' section channelling (generally, steel). In such cases, the tool will not always fit within the flanges and so a barrel extension is required (see Figure 2.28). Such extensions also come into play in some concrete work (though care should be taken that the concrete within the groove or channel is thick enough for using a PA tool).

As with any adaptor or fitting to a PA tool, only those that are specifically designed for the task, and for that particular tool, may be fitted.

Blocking out

When working with 'C' and 'H' section channelling, in particular, and in grooves generally, it is wise to 'block out' the channel. This prevents material being 'channelled' down the metal section and being flung out either end at high velocities (as the ends may be some distance from the operator).

Firing into spalled or damaged surfaces

If considering firing into spalled or damaged surfaces – DON'T!

This simple instruction is one you must adhere to, or risk serious injury or death. A **spalled** surface is one that has been damaged, generally by the failed penetration of a previously fired PA tool fastener. Spalling may also be produced by drilling or the attempted fixing of other mechanical fasteners such as Dynabolts™ and the like.

Spalled surfaces are unstable and dangerous, as the fastener can easily deflect, ricochet and fly in any direction. It is also likely that material from the surface will fly back up at the operator. In addition, any attempt to fix into a spalled surface is almost certain to fail, as there is little chance that the fastener will be able to meld with its surrounding material (Figure 2.27).

Never attempt to fire into a spalled surface.

FIGURE 2.27 Firing into spalled concrete

A summary of dos and don'ts of PA tools

In Appendix 2 at the end of this chapter you will find a brief summary of the dos and don'ts of working with PA tools as outlined in this chapter. Adapted from the safety briefs of two of the world's leading manufacturers and suppliers of PA tools, it acts as a simple reminder that even the most experienced user should never take these tools for granted.

Adaptors and accessories for PA tools

There are so many adaptors and accessories for contemporary PA tools that it is outside the scope of this book to cover them all. Most adaptors, however, fit into one of five categories:

1. barrel extensions
2. fastener guides
3. magazine adaptors
4. extension arms
5. spall guards.

Barrel extensions

These have been covered to some degree in the section titled 'Fixing into grooves and channels' earlier in the chapter. Extensions (Figure 2.28) range in shape and length depending upon the requirements, some allowing for work in grooves as narrow as 15 mm.

Given that virtually all modern PA tools are low-velocity designs, this means that in extending the barrel, the piston must be made correspondingly longer as well. So, in buying a barrel extension, you must also purchase and fit a new, longer piston.

It is vital in fitting or removing a barrel extension that the appropriate piston is installed or reinstalled: too long, and the piston will impact upon the material and perhaps shatter, bend or damage the barrel itself; too short, and the fastener can ricochet back up the barrel, again possibly damaging the tool.

Source: Hilti (Australia) Pty Ltd

FIGURE 2.28 Examples of various barrel extensions and associated pistons

Fastener guides

Depending upon the manufacturer, the term *fastener guide* may cover barrel extensions as well as other adaptors. In other cases, they can be adaptors designed specifically to aid in the positioning of fastenings into particular types of proprietary channel or other materials (such as suspended ceiling components).

Included under this heading are 'disc' adaptors (Figure 2.29). These have been available for some time, and so may be found to fit early direct-acting tools as well as contemporary equipment. A disc adaptor holds a disc, more commonly known as a 'washer', carefully centred on the front of the tool. This allows the operator to safely and accurately fire fasteners through the hole in the washer. Disc adaptors are commonly used to fix light metals, plastics and soft sheets to base materials such as concrete or steel. The disc, or washer, prevents the fastener from simply punching through the material to be fastened.

Source: Hilti (Australia) Pty Ltd

FIGURE 2.29 Examples of a disc adaptor and fastener guide

As with all adaptors, you should read and follow the manufacturer's instructions carefully when fitting and using. Preferably get a representative from the supplier to demonstrate its safe use first.

Magazine adaptors

These adaptors are generally only available for semi-automatic PA tools that use strips of cartridges or charges. These tools already have a space or 'magazine' into which the strips are fed. The magazine adaptor (Figure 2.30) is designed to go on the front of the tool and holds an equal or greater number of fasteners. Now the tool is virtually fully automatic in that it can be fired more-or-less as fast as you can relocate the tool and pull the trigger. (Some tools need cycling prior to the next firing – see 'Cycling the tool' earlier in the chapter.)

Source: Ramset Reid, a Division of ITW Australia Pty Ltd

FIGURE 2.30 Fastener magazine adaptor fitted to a fully-automatic indirect-acting tool

Most magazine adaptors are designed for a limited range of fasteners (some for only one type). As with all adaptors, you must never attempt to fit a magazine to a tool for which it was not designed. Nor should you attempt to fit fasteners into a magazine not designed for them.

Extension arms

The purpose of extension arms (Figure 2.31) is to allow the operator to use the tool overhead (though they can also be used horizontally or downwards). Most adaptors of this type are designed for semi-automatic or fully automatic tools and are seen in use primarily in the fixing of suspended ceiling channels (thereby eliminating the need for an elevating work platform). Extension systems include a trigger system and often are used in conjunction with some form of fastener guide to aid in the positioning of the tool with respect to the material being fastened.

Spall guards

As mentioned previously, spalling is the damage done to a surface (generally used in reference to concrete) by the penetration – or, more generally, the failed

Source: Ramset Reid, a Division of ITW Australia Pty Ltd

FIGURE 2.31 Extension arm in use

penetration of a fastener. The role of the spall guard is to protect the operator from spall (bits of concrete or other material off the work piece) flying back up at them. The risk of flying spall is much greater with direct-acting tools than with their indirect counterparts, and so a guard is always fitted to direct-acting tools.

You must *never* fire a direct-acting tool without the guard correctly positioned.

Note: As stated in earlier sections, when using any adaptor, ensure that it is designed specifically for both the task and the tool. *Never* adapt an adaptor.

New tools and uses

There has been a slow but steady increase in the adaptive use of PA tools over recent years. Hilti, one of the major supplies to Australia, has been particularly active in finding new applications for PA tools in construction, and likewise in developing new ways of adapting them to these new purposes. In the past, all shear studs on concrete and steel composite floors were fixed by welding.

Source: Hilti Corporation, Schaan, Liechtenstein

FIGURE 2.32 A new generation of PA tools; more ergonomic and using digital adjustment of power

The DX 9-ENP Decking Tool claims to be faster, more ergonomic (the operator stands upright rather than bending down using the welded system) and far more precise. Manufacturers, such as KingFlor, still state in their manuals that steel decking is to have welded shear studs; however, it is only a matter of time before this new approach takes over due to its reduced consumption of electrical power.

The digital age has also caught up with PA tools, with some companies offering LED screens and digital readouts for setting the power ratings, piston travel and the like.

Removal of fasteners

Previously, it was explained that PA fasteners tend to 'meld' to some extent with the material into which they have been driven. For this reason, it is virtually impossible to remove them as you would extract a nail or screw from a piece of timber. Attempting to do so is not only likely to damage the surface if 'successful' (particularly in concrete), but is also dangerous. This is because PA fasteners are made of hardened or tempered steel and should they snap or break as you attempt to pull them out, pieces can fly off at very high velocities.

PA fasteners will also fly off at highly dangerous velocities if you deliberately snap them by bending them backwards and forwards (such as by hitting them side to side with a hammer) as you might with a normal nail in timber. This action is prohibited with PA fasteners.

The approved technique for the removal of PA fasteners is in AS/NZS 1873. The tool itself is quite simple and can be made on-site if needs be (Figure 2.33). It is a short length of narrow steel pipe (something just larger than the head of the fastener is appropriate: generally, 10 mm is acceptable) with one end closed off – either by

FIGURE 2.33 Tool for removing fastener

crimping or otherwise squashing flat or by being welded shut. Ensuring that those in close proximity (including yourself) are wearing eye protection, place the pipe over the fastener and rock it from side to side. When the fastener ultimately snaps off, it will be trapped within the pipe and present no risk to you or others.

LEARNING TASK 2.3

1. **Circle 'True' or 'False'.**
   You are allowed to make your own tool for removing fasteners from steel or concrete by using a length of small diameter steel tubing blocked at one end.
   True False
2. **Circle 'True' or 'False'.**
   When changing barrel extensions, you must also change the piston.
   True False

Maintaining PA tools

There is a set schedule for the maintenance of PA tools that is laid out in AS/NZS 1873. To this schedule may be added a variety of additional actions required by the manufacturer (particularly concerning guides and other adaptors). It is critical that you follow this schedule precisely. Evidence of your adherence to the prescribed schedule is recorded in a logbook specific to each tool.

The general schedule of maintenance is as follows:

- Inspect the PA tool for serviceability immediately prior to use.
- Clean and lubricate the tool after each day's use (Figure 2.34).
- Dismantle, inspect, lubricate and check the PA tool for defects weekly or after 60 hours' usage maximum.
- Generally keep the PA tool in safe working order (meaning: well lubricated; dirt- and grit-free; no fractured or damaged parts; all guards in place; and fasteners, charges and the tool itself stored safely – never loaded).

FIGURE 2.34 Maintenance tools necessary for cleaning and lubricating the PA tool

Note: As the trained operator of the tool, you have the right to declare a PA tool unsafe for use.

A PA tool must be completely overhauled by an authorised person (the manufacturer, someone nominated by the manufacturer or a qualified gunsmith) every 12 months. This authorised person issues a certificate upon return of the tool, warranting that such service has been undertaken. As the trained operator, you must enter the service details (date, time and place of overhaul) into a logbook (though this may be done by the authorised person). As mentioned above, a separate logbook is kept for each PA tool you have in service (see Figure 2.35).

Defective PA tools

Should you or another trained operator determine that a tool is unsafe to use, *never attempt to repair it yourself.* Defective PA tools may only be repaired by an authorised person (such as the manufacturer, their nominee or a gunsmith). Likewise, if you believe or are aware that someone other than an authorised person has worked on or attempted to repair a PA tool, you must reject it as still being defective.

Date	Type of service D.W.A.R.*	Hours used	Details of service components used	Signature	Status	Serial number
3/4/24	Weekly	45	Cleaned	D. Smith	OK	25168
3/5/24	Annual	45	Complete overhaul Replaced firing pin (chipped)	G.H. Wilkins	OK	25168

*D: daily; W: weekly; A: annually; R: repair

FIGURE 2.35 Example logbook entries for service details of a PA tool

Clean-up and stowage

Cleaning of your PA tool has been covered to some degree by the previous section on maintenance. The critical aspect of cleaning these tools prior to stowing them away is to ensure that they are unloaded: unloaded before attempting to clean them, and unloaded before placing them back into their lockable storage container (Figure 2.36).

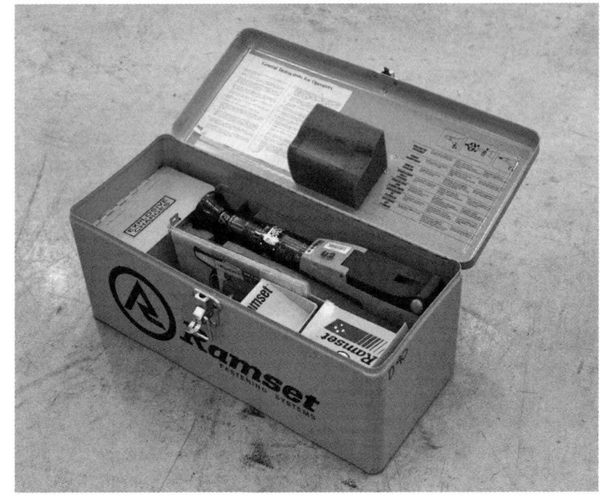

FIGURE 2.36 Check again that the tool is unloaded before placing it back into its lockable storage container

For the process of cleaning your specific PA tool, you will need to follow the supplied manufacturer's guide. In general, it will include the following:

1. Check for and remove any cartridges (live or expended).
2. Disassemble the barrel and remove the piston.
3. Clean the piston and barrel of carbon from exploded charges.
4. Clean the breech (where accessible).
5. Lightly oil with supplied spray.
6. Reassemble and check action (without cartridges).
7. Check again that the tool is not loaded, and then stow.

HOW TO

ALWAYS CHECK!

Yes, you checked before you cleaned; now check again anyway before you stow. People have been known to test fit a cartridge or strip after cleaning and then leave it there ...

So, check!

After cleaning and stowing the tool, you need to clean and disestablish the 'no go' area. Only pack up the 'no go' area barriers and signage after you have completed the following:

- cleaned and stowed the PA tool(s)
- swept the area of all casings, ensuring that no live and/or misfired cartridges are present
- checked that you have stowed all adaptors or other fittings
- checked that you have stowed all unused fasteners
- cleaned and stowed all PPE equipment.

Once this has been completed, the signage may be removed and the area is safe to be opened for general work once more. If working on a major construction site, the site supervisor will need to be informed that the work has been completed and the area is safe for other trades: in general, only the site supervisor may officially reopen the area for other work.

LEARNING TASK 2.4

1. **Circle 'True' or 'False'.**
   The only person allowed to declare a PA tool unsafe for use is a qualified gunsmith.
   True False
2. **The first and last thing you do before cleaning and stowing a PA tool is to:**
   a check that you have removed all fasteners
   b check that you have correctly removed all adaptors and fittings

GREEN TIP

Make sure you follow safe practices for disposal of spent explosive power tool cartridges that have fired or not fired. You can cause a lot of injury or damage to the environment by having unexploded cartridges moving through the waste chain from your building site.

Make sure that when you use an explosive PA tool you have been trained by a competent person and you are very familiar with the tool's user manual so that you know what is safe to do and what is not safe to do. Don't become overconfident with a dangerous tool like this.

COMPLETE WORKSHEET 3

SUMMARY

Some final words of caution.

The risk when you are a new user of PA tools is that initially you are a bit wary – maybe even a little scared – about using them. However, given time and some level of experience, you can soon begin to become over-confident due to the simplicity of the tool and the fact that nothing 'bad' has happened. This is where you must take stock and reflect upon your training. Bad things can and do happen: that's why the training in the use of these tools is so intense. Bad things happen when we least expect it, and this is generally when we have let down our guard and neglected to act upon some very basic aspect of the training – such as keeping the tool perpendicular to the surface, using the lowest charge first in testing, or checking that the tool is unloaded before inspecting, cleaning or stowing it away.

The basics of PA use were detailed in each of the main sections of the chapter, and may be summarised as follows:

- Older tools used a direct action approach, which meant the fastener travelled faster and further if by accident it passed through the material it was fired into.
- By regulation, the provision of training by employers on each specific tool is mandatory. In addition, a SWMS or JSA must be developed for each context in which the tool is to be used.
- Being able to identify the difference between direct-acting and indirect-acting PA tools is vital for your safety and that of those around you, due to the vastly greater power and muzzle velocity of the former.
- Be knowledgeable about which materials you can and cannot fire a fastener into or through. Some materials are too brittle, while others are too hard, soft or thin. Fastener and charge selection is also critical to ensure appropriate penetration into or through the base material.
- For the safe use of PA tools, establish a 'no go' area around the work location and use appropriate signage. Check for flammable gases, compressed air and the like. Test the base material for suitability, and check for electrical wiring, gas or fuel pipes that may be behind or embedded in that material. Check that you have all the appropriate PPE and that the tool is safe to use, following the manufacturer's pre-firing instructions.
- Only load where and when you are to use the tool, following the manufacturer's instructions. Make sure that you never point the tool, loaded or unloaded, towards yourself or anyone else.
- When firing the PA tool, follow the manufacturer's instructions, making sure that you have good communication with your spotter(s) and that you loudly call, 'Firing', moments prior to pulling the trigger. Ensure you understand the misfire procedure, and follow it should a misfire occur.
- Only use an adaptor made specifically for that tool and where it is designed for the purpose you intend.
- Follow the manufacturer's recommendations for frequency of maintenance, ensuring inspection prior to and immediately after use of a PA tool. Check the tool's logbook and make sure all required inspections have been completed.
- Additional to the general clean-up, you must ensure that no spent, live or misfired cartridges are left lying around, and that all adaptors and other fittings are stowed away.

If you keep your wits about you and follow the standard procedures, your PA tools will give you many years of safe and useful work.

REFERENCES AND FURTHER READING

Text

Frank M, et al. (2012), *Ballistic parameters and trauma potential of direct-acting, powder-actuated fastening tools (nail guns)*, **http://www.ncbi.nlm.nih.gov/pubmed/21607714**

Workplace Health and Safety Australia, **http://www.healthsafety.com.au/safe-work-method-statementsswms/**

Resources

Hilti Corporation, phone: 131 292, **http://www.hilti.com.au**

Hilti DX 351 User Manual: **https://www.hilti.com.au/medias/sys_master/documents/he4/9083527757854/DX_351_EN_PUB_5126242_000.pdf**

Ramset, phone: 1300 780 063, **http://www.ramset.com.au**

Relevant Australian Standards
AS/NZS 1873.1 Powder-actuated (PA) hand-held fastening tools – Selection, operation and maintenance
AS/NZS 1873.2 Powder-actuated (PA) hand-held fastening tools – Design and construction
AS/NZS 1873.3 Powder-actuated (PA) hand-held fastening tools – Charges
AS/NZS 1873.4 Powder-actuated (PA) hand-held fastening tools – Fasteners
Note: At the time of writing, the above four standards show as having been withdrawn. However, as nothing has replaced them, their use is still encouraged until a new set of standards becomes current.
Note: Withdrawn standards are still able to be called upon by governments and industries as required. Withdrawn does not mean superseded or obsolete. Standards are withdrawn after a period of time in which calls for their review have not been acted upon.
AS/NZS 1270 Acoustics – Hearing protectors
AS/NZS 1337.0(Int) Personal eye protection – Eye and face protectors – Vocabulary
AS/NZS 1337.1 Personal eye protection – Eye and face protectors for occupational applications
AS/NZS 2210.3 Class 1: Protective footwear

APPENDIX 1

STEPS IN COMPLETING A SAFE WORK METHOD STATEMENT (SWMS)

1 **Identify** any *hazard or potential hazards* (see Table A.1).
2 **Assess** the level of risk and determine priority using a *risk management matrix* (Table A.2).
3 **Control** the risk using the *hierarchy of control* (Table A.3).
4 **Review** and **evaluate** both the *risks and the control measures* on a continual basis (Figures A.1 and A.2).

TABLE A.1 Examples of potential hazards

Note: Examples of potential hazards may include, but are not limited to, those shown in the table.				
Electrical	**Mechanical**	**Pressure**	**Chemical**	**Gravity**
Contact with live wires or terminals causing shock, flash to eyes, burns or falls. Discharge of capacitors causing same.	Caught by operating machinery. Struck by moving machinery or objects. Caught by movement of mechanical parts. Crushed by objects moving or falling. Entrapment. Excessive vibration. Flying spall.	Injury from release of stored energy in hydraulics, pneumatic springs. Recoil from PA tools.	Fire or explosion from build-up of flammable gases. Ignition of existing flammable products, contaminants/toxins, causing suffocation, burns or poisoning. Gases from PA tools used in confined space.	Falls from or into vessels. Falls from structures. Impact injuries from falling objects. Engulfment by product/materials. Working overhead with heavy tools.
Noise	**Radiation**	**Biomechanical**	**Biological**	
Sound levels > 85dBA causing hearing damage from machinery or equipment. Explosive discharge from PA tools.	Extremes of temperature. Burns. UV from welding flashes. UV from exposure to sun. X-ray exposure. Eye damage from lasers.	Strains and sprains from lifting objects. Strains and sprains from moving objects. Slips and trips from spillage/slippery surfaces. Uneven/unstable surfaces. Poor lighting. Crush injury.	Disease or illness from spores; e.g. legionnaires' disease from infected blood products, hepatitis, brucellosis.	

TABLE A.2 Risk management matrix

Consequence: ***How SEVERELY could it hurt someone?*** ***Or*** ***How ILL could it make someone?***		***Likelihood: How likely is it to be that bad – what is the probability of it happening?***			
		VERY LIKELY **Could happen anytime** **++**	**LIKELY** **Could happen sometime** **+**	**UNLIKELY** **Could happen very rarely** **–**	**VERY UNLIKELY** **Probably never will happen** **– –**
KILLED OR PERMANENTLY DISABLED	X	1	1	2	3
LONG-TERM ILLNESS OR SERIOUS INJURY	!!!	1	2	3	4
MEDICAL ATTENTION LOST TIME DUE TO INJURY	!!	2	3	4	5
FIRST AID	!	3	4	5	6

TABLE A.3 Hierarchy of control

Elimination	Can we do without? Is this action really necessary?
Substitution	Can we use something less dangerous, or carry out a different action?
Isolation	Can we enclose, fence off, or otherwise create a ‘no go’ and/or exclusion zone?
Engineering	Can we use protective guards or venting, or redesign tools or equipment?
Administration	Can we develop safer processes, procedures, monitoring and review systems, or training?
Use of PPE	Can we develop training in the correct use of PPE?

Company/Contractor: Adventure Construct		**Date:**			**Permit to work required:** **Yes No**	**RA/SWMS no.**
Site name or principal's name: G. Bean					**Site address: Civilisation Gate, Hallet**	
Name of person completing form: G.H. Wilkins					**Signature: G.H. Wilkins**	
Operating a powder-actuated tool						
Activity: List in order the tasks required to complete the work activity	**Hazards:** List the hazards that are present in each task	Severity	Likelihood	Priority	**Risk control methods:** List the methods to be used to eliminate or minimise the risk of injuries arising from the identified hazard	**Who is responsible:** Write the name of the person who is responsible for ensuring the control measure is implemented
Establish 'no go' area	Manual handling	!	–	6	Inspect work area prior to importing equipment. Ensure correct lifting techniques	**G.W.**
Inspect PA tools	Loaded tool	X	–	3	Only trained operative to inspect	**G.W.**
Inspect base material	Ricochet	!!	–	3	Eye protection, trained operator only	**G.W.**
Load/test fire	Ricochet, over-penetration	X	–	2	Full PPE, trained operator only, spotter in place	**G.W.**
Fixing	Ricochet	X	–	2	Full PPE, trained operator only, spotter in place	**G.W.**
Clean-up/stowage	Loaded tool. Manual handling	X	–	3	Only trained operative to inspect and stow. Ensure correct lifting techniques	**G.W.**

FIGURE A.1 Example risk assessment and safe work method statement (SWMS): Part 1

Personal qualifications or experience required to undertake activities	***Personnel duties and responsibilities of those undertaking activities***	***Any training required to undertake activities***
Be over the age of 18 and be trained generally and specifically in PA tool use: Hilti DX 460; Ramset D70	As per Part 1 of this SWMS	Site induction

Engineering details/Certificates of competence approvals required before tasks can be undertaken	**Legislation to be complied with when undertaking activities**
SDS for PA tools cleaning products and charges	State-applicable WHS/OHS Act and Regulations
PA manufacturer's technical information (tool, fasteners, charges)	State-applicable WHS/OHS Act and Regulations Manufacturer's User's Manual
AS/NZS 1873.1 Powder-actuated (PA) hand-held fastening tools – Selection, operation and maintenance AS/NZS 1873.2 Powder-actuated (PA) hand-held fastening tools – Design and construction AS/NZS 1873.3 Powder-actuated (PA) hand-held fastening tools – Charges AS/NZS 1873.4 Powder-actuated (PA) hand-held fastening tools – Fasteners AS/NZS 1270 Acoustics – Hearing protectors AS/NZS 1337.0(Int) Personal eye protection – Eye and face protectors – Vocabulary AS/NZS 1337.1 Personal eye protection – Eye and face protectors for occupational applications AS/NZS 2210.3 Class 1: Protective footwear	State-applicable WHS/OHS Act and Regulations Manufacturer's User's Manual

Plant and equipment required to complete the tasks	**Pre-operation and/or maintenance checks to be completed before tasks are to be undertaken**
Hilti DX 460	Yes: Full pre-operational check before use
Ramset D70	Yes: End-of-day check and logbook entries

Read and signed by all persons undertaking activities stated on this document			
Print name	**Signature**	**Date**	**Supervisor/Principal/Employer signature**
Frank Hurley	*Frank Hurley*	13/4/2020	*George Bean*
Douglas Mawson	*D. Mawson*	13/4/2020	*George Bean*

FIGURE A.2 Example risk assessment and safe work method statement (SWMS): Part 2

APPENDIX 2

BASIC SAFETY STEPS IN THE USE OF PA TOOLS

1 NEVER attempt to use a PA tool UNLESS you have satisfactorily completed authorised general training and training on that specific PA tool.
2 ALWAYS read and ensure you have understood the instruction manual before using.
3 When not in use, the PA tool should be placed, unloaded, in its box or case, which should then be shut, locked and stored in a safe place to which only authorised persons have access.
4 ALWAYS follow the directions for care of the tool contained in the manual.
5 ALWAYS fit and use special adaptors strictly in accordance with the supplied manuals and/or directions of the supplier.
6 NEVER adapt or modify an adaptor to suit another tool, fastener or application.
7 ALWAYS display approved warning notices so they are clearly visible to all persons who are at, or near, the place where a tool is being used.
8 NEVER fire the tool unless you and any other people in the area are wearing approved eye and hearing protection.
9 NEVER use PA tools near explosive or inflammable gases, dusts or vapours in a compressed atmosphere.
10 NEVER bring a PA tool or its charges into a hot atmosphere, as charges may unintentionally explode or become dangerous.
11 NEVER leave a tool unattended.
12 NEVER load the tool until ready to fasten.
13 ALWAYS unload a tool not being used immediately.
14 NEVER carry a PA tool around in a loaded condition.
15 NEVER carry fasteners or other metal objects in the same pocket or package as charges.
16 ALWAYS check that the barrel is free from obstructions before loading.
17 NEVER point the barrel of a loaded or unloaded tool towards any person.
18 ALWAYS point the tool downwards and as far as practicable away from the operator's body. This applies in particular where a charge that has misfired is being removed.
19 NEVER place your hand over the muzzle with a charge in the tool.
20 Fasteners should not be driven into brick, concrete or similar substances if:
   a they are nearer than 75 mm to an edge, or
   b the substance is less than 100 mm thick, or less than three times the shank penetration into the base material, whichever is the lesser, or
   c they are nearer than 150 mm to where a former fastener has failed.
21 Fasteners should not be driven into steel if:
   a they are nearer than 12 mm to an edge or hole
   b the steel is less than 4 mm thick, unless, for fastening into thinner steel, a specialised fastener is used.
22 NEVER fasten into wood, fibreboard, plaster or other soft materials unless backed up by a material that will prevent the fastener from passing completely through (masonry, concrete or thick steel).
23 NEVER fire fasteners into brittle or hard materials such as glazed brick or tile, terracotta, marble, granite, slate, glass or hardened steel. Do not attempt to fasten into high-tensile steel, cast iron, heat-treated steel, or pressurised vessels such as gas bottles, compressed air cylinders, etc. Take great care when test firing in concrete older than three years.
24 NEVER attempt to fasten into a spalled area in masonry or concrete where a previous fastener has failed.
25 NEVER fire fasteners through existing holes unless the tool is fitted with a specifically designed adaptor for accurate alignment of the barrel.
26 DO NOT fire into material that dulls the point of a test fastener that has been hit into it by means of a hand hammer.
27 ALWAYS position your body comfortably in line with and behind the tool.
28 ALWAYS ensure the tool is at right angles (90 degrees) in both directions to the work surface before firing.
29 NEVER fire the tool at an angle to the work surface.

30 NEVER use the tool at or near the extremities of your reach.
31 NEVER use PA tool charges or cartridges in firearms.
32 ALWAYS wear safety eye and hearing protection when using a PA tool.

Misfires. The tool is to be kept firmly against the work surface for 10 seconds; the same charge shall then be fired again. If it fails the second time, count again to 10 and then remove the tool, keeping it pointing away from you or others. Remove the charge, and place in water if possible (not with strip charges). Retain the charge in a safe place for collection and disposal by the supplier.

Jammed and loaded PA tools. Should a PA tool jam in the firing position, it must be locked in a secure place. This location should be where no harm could result if the tool were to discharge. *DO NOT attempt to fix or adjust the tool yourself.* Jammed PA tools may only be serviced by an authorised person (i.e. the manufacturer, their nominee or a qualified gunsmith).

GET IT RIGHT

FIGURE 2.37 Creating a 'no go' zone

In Figure 2.37, workers have attempted to establish a 'no go' zone in a public space. Based on the information contained in the chapter, describe how this may be improved and what is missing for the safe use of PA tools.

___

___

___

WORKSHEET 1

Student name: ______________________

Enrolment year: ______________________

Class code: ______________________

Competency name/Number: ______________________

To be completed by teachers	
Student competent	☐
Student not yet competent	☐

CPCCCM2007: Use explosive power tools and CPCCCA3027: Set up, operate and maintain indirect action powder-actuated power tools – specifically, the elements 'Plan and prepare' and 'Set out fasteners'.

Task: Answer the following questions.

1 What are the four Australian Standards that apply specifically to PA tools?

1 ______________________

2 ______________________

3 ______________________

4 ______________________

2 What is the authority (in your state/territory) that issues further guidance, compliance codes and the like?

____

____

____

3 Why is it that we have changed the name from explosive power tools to powder-actuated tools?

____

____

4 List five things a PA tool is designed to do.

1 ______________________

2 ______________________

3 ______________________

4 ______________________

5 ______________________

5 Circle 'True' or 'False'.

Direct-acting PA tools are safer as they have a lower velocity than indirect-acting PA tools.	True	False

6 Mild steel and concrete less than three years of age are both base materials that are appropriate to fire a fastener into using a PA tool. What is the third approved base material and what are the cautions around firing into it?

____

7 Which of the following materials must you never fire into with a PA tool?

a Glass

b Cast iron

c High tensile steel

d All of the above

8 Why can the likes of plasterboard, timber or thin steel be fired 'through', but not be used as a base material?

____

____

9 Circle 'True' or 'False'.

You can fix into 3 mm and 4 mm steel only when using special fasteners designed for that purpose and that specific tool.	True	False

10 What is the cartridge type used in all modern PA tools?

____

____

11 Label all the parts of the indirect-acting tool shown below.

Source: Hilti (Australia) Pty Ltd

12 PA tools use what are known as 'rim fire' cartridges. When one of these cartridges misfires, the following has occurred:

a The cartridge casing has split and the grains of gunpowder have drifted into the barrel

b The primer in the rim has partially ignited and is burning slowly

c The primer in the rim has failed to ignite at all

d Either 'b' or 'c' may be occurring

13 Fill out the table below stating the colour of each of the various strengths of charges available.

Strength	Colour
Minimum	
Very weak	
Weak	
Medium	
Medium strong	
Strong	
Very strong	
Especially strong	
Maximum	

14 List the four strengths and colours of charges commonly available as strips and used in modern semi-automatic tools.

1 ______________________________

2 ______________________________

3 ______________________________

4 ______________________________

15 State the two main requirements of the container used for storing cartridges.

1 ______________________________

____

2 ______________________________

____

16 The text lists seven things you must never do with charges. Two of these are:

Never leave live charges lying around.

Never use a PA tool charge in a firearm.

List the other five:

1 ____________________

2 ____________________

3 ____________________

4 ____________________

5 ____________________

WORKSHEET 2

To be completed by teachers	
Student competent	☐
Student not yet competent	☐

Student name: ______________________

Enrolment year: ______________________

Class code: ______________________

Competency name/Number: ______________________

CPCCCM2007: Use explosive power tools and CPCCCA3027: Set up, operate and maintain indirect action powder-actuated power tools – specifically, the elements 'Set out fasteners', 'Prepare powder-actuated power tool' and 'Operate powder-actuated power tool'.

Task: Answer the following questions.

1 A 'no go' area as it applies to PA tools is:

  a An area near where you are working that you should store cartridges and other dangerous equipment
  b A clearly defined area into which people, other than those required to complete the task(s) involved, are restricted from entering
  c Only required when there is likely to be pedestrian traffic nearby
  d Always a minimum of 6 m in radius

2 Describe the wording, colour and size of the PA tool warning sign.

____

____

____

____

3 A spotter is someone who:

  a Tells you when and where to fire the next fastener
  b Maintains the PA tool
  c Checks the PA tool prior to use
  d Someone who checks the area for safety prior to firing

4 List the three things you must check before you bring the PA tool to the firing location.

  1 ______________________
  2 ______________________
  3 ______________________

5 Describe the procedure for checking if a material is suitable for firing into.

6 With concrete that is over three years old, who should you ask if it is suitable for firing a PA tool into?

7 Base material such as precast concrete walls may have hidden services such as gas lines or electrical wiring. What are three ways you can identify these risks?

1

2

3

8 When firing into a base material that might just allow a fastener to fly right through it, one possible option is to:

a Set up a 'no go' safety area behind the material

b Use the lowest charge and build your way up to save having to use a spotter

c Test fire one shot and then go see what happened

d Fire through a sheet of plywood to reduce the penetration power

9 Before firing a PA tool, what checks must be made of the tool itself?

10 What PPE *must* be worn by operators of PA tools?

11 Aside from the manufacturer's instructions for a specific tool, list the five basic actions governing the loading of a PA tool.

1 ____

2 ____

3 ____

4 ____

5 ____

12 Describe the loading and cycling sequence.

____

____

____

____

____

____

13 Circle 'True' or 'False'.

Most single shot or semi-automatic PA tools require that you 'cycle' the tool prior to firing. You should never walk around with a tool that has been cycled and therefore loaded ready to fire.	True	False

14 What must your spotter or assistant call out prior to you firing a shot?

____

15 What must the operator of a PA tool call out prior to firing a shot?

____

16 Circle 'True' or 'False'.

Generally, when using steel as the base material, a fastener should protrude approximately 6 mm out the back to be considered correctly driven.	True	False

17 From the information below, determine the appropriate length of a threaded stud.
Shank diameter is 3.8 mm.
Timber to be fixed is 45 mm thick.
Fixing into 6 mm steel.

a 32 mm

b 50 mm

c 57 mm

d 65 mm

18 Describe how you can reduce the power of both direct- and indirect-acting PA tools without having to use a lower-powered charge.

Direct-acting tool: ______________________________

____

Indirect-acting tool: ______________________________

____

19 Outline the misfire procedure for both single-shot and semi-automatic tools.

Single-shot tool: ______________________________

____

Semi-automatic tool: ______________________________

____

20 Fill out the table below, with regard to the spacing and edge distances for fasteners.

Fastener spacing and edge distances		
Material	**Spacing (mm)**	**Distance from edge (mm)**
Steel		
Concrete		

21 A spalled surface:

a Must never be fired into

b Can only be fired into when using a spall guard

c Is created when using a spall guard

d May be identified by heavy, flaking rust on steel

22 When is blocking out required, and why?

____

____

____

____

WORKSHEET 3

Student name: ______________________

Enrolment year: ______________________

Class code: ______________________

Competency name/Number: ______________________

To be completed by teachers	
Student competent	☐
Student not yet competent	☐

WORKSHEETS 2

CPCCCM2007: Use explosive power tools and CPCCCA3027: Set up, operate and maintain indirect action powder-actuated power tools – specifically, the elements 'Maintain powder-actuated power tools and kits', 'Store powder-actuated power tools and charges' and 'Clean up'.

Task: Answer the following questions.

1 What is a disc adaptor? Why is it necessary?

What:

____

Why:

____

2 When might you use an extension arm?

____

____

____

3 When fitting a barrel extension to an indirect-acting PA tool, you must also fit what else to it? State why this is necessary.

What: ______________________

____

Why: ______________________

____

4 Within the chapter, a particular tool and method of fastener removal is described. Why is it necessary to do it in this manner?

____

____

____

____

____

5 State the two main requirements of the container used for storing cartridges.

1 ______________________________

____

2 ______________________________

____

6 Before and after cleaning your PA tool, you must always:

a Ensure no one else is on the premises
b Disassemble and reassemble the tool
c Spray a light coating of oil into the barrel of the tool
d Check that the tool is not loaded with any cartridges

7 What five things must you do before disassembling the 'no go' area?

1 ______________________________

2 ______________________________

3 ______________________________

4 ______________________________

5 ______________________________

8 When you have finished your PA work on a major construction site, who should you inform that you have completed the work?

a Your company's owner
b The site security personnel
c The site supervisor
d One of the other trades that were working nearby

9 Defective PA tools may be repaired by:

a A trained PA tool operator
b The agent who sold you the tool
c The manufacturer, their nominee, or a gunsmith
d The spotter under the guidance of a trained PA tool operator

10 Circle 'True' or 'False'.

As a trained PA operator you are able to replace pistons, barrels and associated clips.	True	False

11 What is the general schedule of maintenance for PA tools?

___

___

___

___

12 Every 12 months the PA tool must be completely overhauled. Who may perform this work?

___

13 Who may declare a PA tool to be defective?

___

___

14 Before cleaning a PA tool, and before placing it in storage, what must you always do?

___

___

ELEVATING WORK PLATFORMS

3

This chapter covers the following topics from the competency 'Elevating work platforms':

1 Background
2 Types and designs of EWPs
3 Logbooks and manuals
4 Safety around hydraulics
5 The safe use of EWPs: Part 1 – spotters, hazards and controls
6 The safe use of EWPs: Part 2 – checks, loads and positioning

Chapter 1 of this text began by suggesting that most tradespeople working in the construction industry will need to do so 'at heights' at least occasionally, if not frequently. Of the many means of accessing high work areas discussed in that chapter, two were identified as needing much deeper enquiry: elevating work platforms (EWPs) and scaffolds (which are addressed in Chapter 4). It is EWPs and their safe use that is the focus of this chapter.

Overview

This chapter addresses the key elements of the following unit:

- CPCCCM3001: Operate elevated work platforms up to 11 metres

Prerequisites

None

Important note: Undertaking this unit does not lead to a High Risk Work Licence for operation of boom-type elevating work platforms with a working height or horizontal reach of 11 m or greater.

Background

The safe operation of EWPs requires an extensive knowledge of the equipment, procedures and regulations, coupled with experience of the many contexts in which they may be used. In addition, there are many and varied forms of EWPs available, each designed for a particular range of purposes and having, as does any means of accessing high work areas, their own inherent risks.

The purpose of this chapter is to provide you with focused resource materials on the regulations and safe practices and procedures applicable to EWPs. In addition, the chapter provides examples of the various equipment types and their appropriate use in the workplace.

Note: Only an experienced operator can teach what you need to know to safely use these machines. Guided instruction and supervised experience with the equipment in the workplace will be required before you can safely and efficiently use these machines on your own.

The national competency CPCCCM3001: Operate elevated work platforms up to 11 metres has been used as a guide in determining the range of equipment and regulatory information covered within the chapter. Where necessary, this range has been extended to ensure that the relevant underpinning knowledge has also been covered.

Types and designs of EWPs

Before delving into the regulations and procedures governing the use of EWPs, it is worth having a first look at the equipment itself. In so doing, some of the key components and the general language surrounding their use can be introduced.

Following AS 2550.10 Cranes, hoists and winches – Safe use – Mobile elevating work platforms, an EWP (or MEWP – mobile EWP) may be defined as:

A relocatable machine or device designed to deliver people, tools and/or material to an elevated area of work, such a device consisting of at least a suitable work platform and some type of extending structure stemming from an appropriately counterweighted chassis. In addition, full positioning controls must be located on the work platform, with a second set accessible at ground level via the chassis.

Note 1: The term *mobile elevating work platforms* does not mean that the equipment must be 'self-propelled'. Trailer-mounted EWPs are still 'mobile', despite their needing to be towed into position.

Note 2: Travel towers are specifically excluded from the AS 2550.10 definition of EWPs.

Types of EWPs

As you can see from the definition, the term elevating work platform (also known as an 'elevated work platform') covers a wide range of equipment capable of lifting (elevating) materials and workers to heights while providing a safe platform to work from once there.

AS 2550.10 clusters these various types into five categories:

1. trailer-mounted boom lifts (TL)
2. self-propelled boom lifts (BL)
3. vertical lifts (VL)
4. scissor lifts (SL)
5. truck-mounted boom lifts (TM).

The type of EWP you need will depend on the job that needs to be completed. To find this information it's important that you determine the job requirements and that the job is planned so you can ensure the correct EWP is selected to complete the work.

Tools and materials required for the task can be selected and the size and weight of this will also play a factor in the selection of the EWP. As always, tools and materials must be checked for any visible signs of damage before using and reported if identified. This includes reviewing the safety data sheets (SDS) for any hazardous materials that are being used during the works.

Each EWP should have its own specific safe work method statement (SWMS) that should also be reviewed by all people involved in the work when making the decision on the best EWP for the job.

Boom-type elevating work platforms

Of the above five categories, three may collectively be referred to as boom-type elevating work platforms. Boom-type EWPs fall under the current work/occupational health and safety (WHS/OHS) legislation as requiring a High Risk Work Licence when a boom-type EWP is capable of positioning:

- the floor of the platform equal to or greater than 11 m above the surface upon which the EWP is set up
- the outer edge of the platform equal to or greater than 11 m from the centre of the EWP when measured horizontally (**Figure 3.1**).

This requirement is applicable to all Australian states and territories. The only exemption is when the EWP is being loaded or offloaded by a truck driver for the purposes of delivery or transport.

Note 1: There are currently no height limitations on vertical or scissor lifts.

Note 2: A High Risk Work Licence for boom-type lifts does not exempt you from the need to have a certificate of competency (not a licence) covering vertical and scissor lifts. This applies no matter their height capacity.

KEY POINTS

- There are five basic types of EWPs.
- However, the important distinction is between boom-type and non-boom-type EWPs – that is, between those with booms and those that use scissor or vertical extension arms.
- Boom-types that can reach over 11 m require a licence to operate.

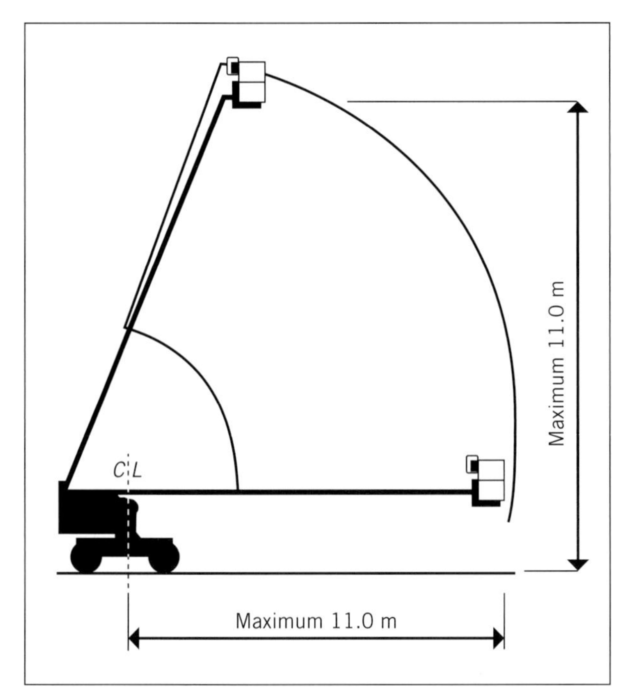

FIGURE 3.1 A boom-type EWP's maximum reach limit before an operator licence is required

Components of an EWP

For the purposes of familiarity and identification, we will look briefly at each of the five EWP types in the following pages. Before doing so though, it is useful to know some of the basic component terminology concerning EWPs generally: take a moment to familiarise yourself with the various components shown in Figure 3.2:

- basket or platform
- boom (upper and lower)
- fly boom (a section that 'telescopes' out of the upper boom)
- hydraulic piston or cylinder
- control module (base and platform)
- turntable
- stabilisers or outriggers
- knuckle or upright
- jib (Figure 3.2) or fly jib (see Figure 3.6).

Given that Figure 3.2 shows a trailer-mounted boom lift, we will look in detail at this type of EWP first.

Trailer-mounted boom lifts

Figure 3.2 shows a typical trailer-mounted EWP extended and work ready. Mounted on a movable trailer, these machines are designed to be towed by any vehicle with the appropriate towing hitch and vehicular mass. Therefore, while still considered to be a mobile EWP, they are not self-propelled.

A key characteristic of these units is their use of **outriggers** or **stabilisers**. It is by this means that an otherwise quite light EWP can extend to significant **working heights** while still remaining sufficiently stable.

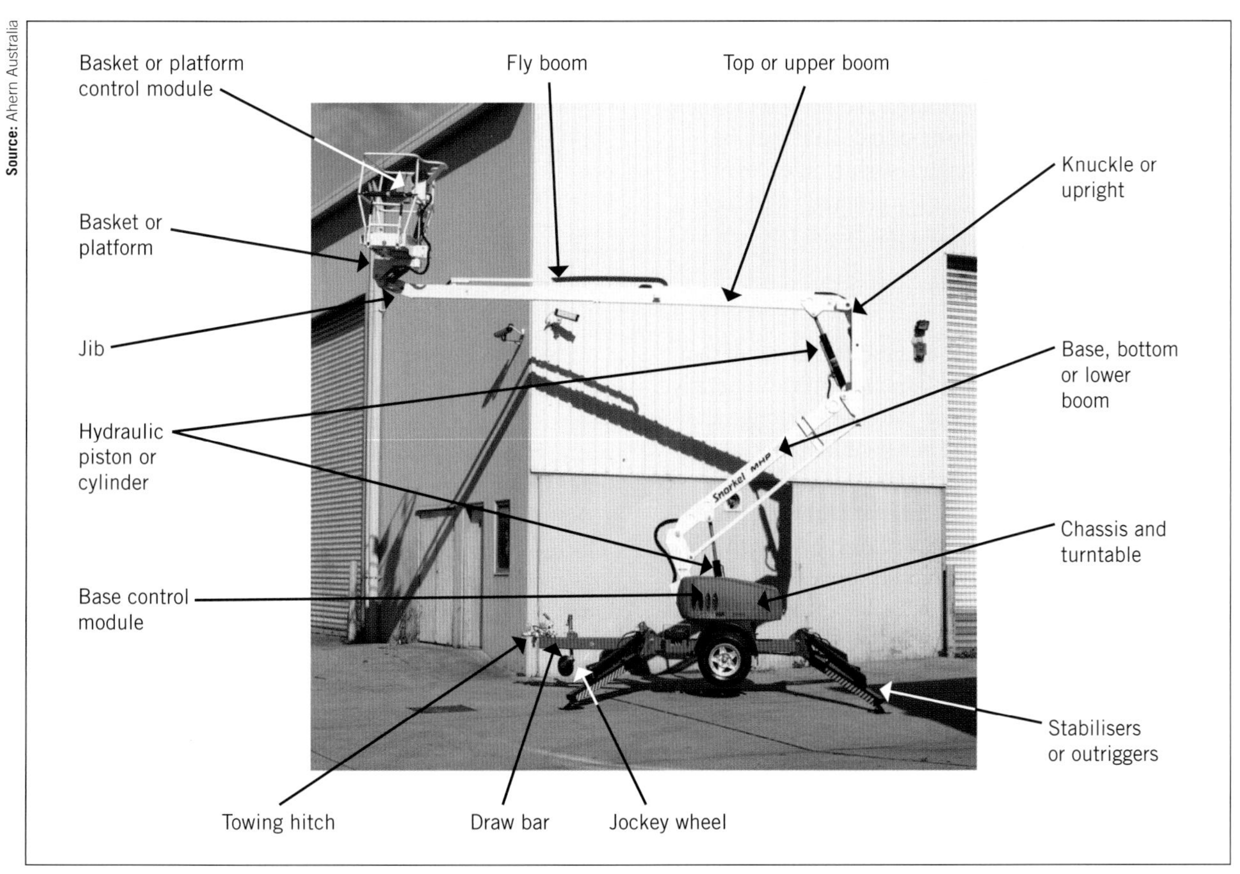

FIGURE 3.2 Trailer-mounted boom lift

Stability is a relative thing, however; and of all the EWPs available, the trailer-mounted units have the most noticeable flex and bounce. As always, careful attention must be given to the maximum load and reach specifications stipulated for the specific machine in use.

ON-SITE

AUSTRALIAN TOWING REGULATIONS

Standard Australian towing regulations stipulate that:

When towing an item *with* brakes, the towed item must *not weigh more than 1.5 times* the weight of the towing vehicle.

OR

When towing an item without brakes, the towed item must weigh less than the towing vehicle.

Note: If the towing vehicle's manufacturer has stipulated a lower maximum towing capacity, or the fitted tow bar has a lower rating, then this lower figure must be adhered to.

It is these same stabilisers that offer one of the main benefits of using trailer-mounted EWPs. Unlike many self-propelled units, these EWPs can be easily set up on sloping or uneven ground. Even slightly unstable ground (such as sand or loam) can be catered for by considered use of sole or 'road' plates, though careful attention must be given to the positioning of outriggers to ensure they all share the load equally (see 'Outriggers, road plates and pigsty packing' later in the chapter).

Despite the limitations of the lightweight components (booms and turntable, compared to larger self-propelled units), trailer-mounted EWPs can be designed to safely reach heights slightly in excess of 25 m. Due to the stabilisers, they generally also have a horizontal **platform reach** of at least 50–60 per cent of their maximum height.

Self-propelled boom lifts

Self-propelled boom lifts are of a much heavier design and construction than trailer-mounted units. Generally, they do not have outriggers or stabilisers (though they can do), and are used where constant mobility is required. Mounted on a self-driven four-wheeled chassis, they are designed mainly for fairly level and stable ground. This surface should also be reasonably smooth, though rough-terrain units are also available.

As with trailer-mounted EWPs, self-propelled boom lifts come with various vertical and horizontal reach capacities. In addition, there are numerous variants, including track or skid-steer units (which use rubber or metal tracks, like those on bulldozers, instead of wheels), rough-terrain and confined-space types (small-bodied units, the rear of which do not swing over the wheels), and machines limited to indoor use only. Further, they may be powered by gas, petrol, diesel or electric motors.

Despite this array, boom lifts can be divided into two distinct types:

1 telescoping or stick boom (**Figure 3.3**)
2 telescoping knuckle boom (refer back to **Figure 3.2**) – also known as an 'articulated boom'.

Source: Ahern Australia

FIGURE 3.3 Self-propelled stick boom lift

The second type, the knuckle or articulated boom, resembles the boom assembly on the trailer-mounted units discussed earlier. By having a lower boom and then a 'knuckle', the **operator** can reach 'over' obstacles as well as gain height. As with many trailer-mounted EWPs, the upper boom (and, on occasion, the lower boom also) is fitted with a fly boom that telescopes (gets longer), thereby increasing the reach. Horizontal reach on these units is limited due to the knuckle array, often being confined to no more than 60 per cent of their elevation capacity.

The first type also has a telescoping upper boom, though generally it consists of multiple 'fly boom' sections. This offers similar height capacity to the knuckle boom, but with greater stability. Without a knuckle, stick booms have a much greater horizontal reach (close to 100% of their elevation reach) than those with a knuckle, which is one reason you may choose to use them. The other advantage that telescoping boom lifts have over knuckle booms is their cost. Being simpler in design, they are cheaper and easier to maintain, making them the obvious choice if height or reach only is required (i.e. they are not required to work over obstacles).

The main advantage of self-propelled boom lifts over trailer-mounted EWPs is their mobility. On large concrete slabs or prepared ground, self-propelled boom lifts can be driven by the operator relatively quickly and easily from one workstation to the next. To aid their manoeuvrability in confined spaces, some machines have both front- and rear-wheel steer. (*Note:* Depending upon the manufacturer's specifications, self-propelled boom lifts may only be driven at 'creep' speed when the platform is elevated, or in some cases may need to be lowered completely prior to relocating.)

Unless fitted with outriggers (or using tracks, as do skid-steer units – which normally have outriggers anyway), all self-propelled boom lifts will be fitted with some form of solid tyres. This limits sway of the extended booms due to tyre flex; most importantly, it also ensures that the machine won't undergo a change in balance due to slow or sudden deflation of the tyres. Rough-terrain machines will generally have foam-fill, tractor-like tyres. Indoor types, or those for use on smooth surfaces such as concrete slabs, will have solid, non-marking tyres not dissimilar to those found on forklifts.

Scissor lifts

Scissor lifts (Figure 3.4) are another, but very different, form of self-propelled EWP. Unlike boom lifts, scissor lifts work only vertically, doing so by means of a scissor-like array of arms raised with hydraulic pistons or cylinders. Like most self-propelled EWPs, they are most commonly designed for stable, flat ground, or concrete slabs.

Source: Ahern Australia

FIGURE 3.4 Scissor lift

Once again, as with boom lifts, there are multiple variations of scissor lifts, including rough-terrain units that will be fitted (generally) with outriggers. These can compensate, to a small degree, for some variations in the ground. Generally, however, the stabilisers on these EWPs are designed to reduce sway on those units with extended reach, some scissor lifts being capable of reaching heights of over 30 m. Only scissor lifts with outriggers may be deployed on anything other than level surfaces.

Other types of scissor lifts include units with extendable decks. These may extend longitudinally (the long length gets longer – Figure 3.5) or out to the side. Before deploying (extending) decks, be sure that you have read and understood any additional load restrictions that may apply – for example, a reduced working load may apply to the extended deck area.

Source: Ahern Australia

FIGURE 3.5 Scissor lifts include units with extendable decks, which extend longitudinally or out to the side

As with boom lifts, the tyres on scissor lifts must be of the solid type. A similar array of power plants is available: gas, petrol, diesel or electric.

Unlike most other types of EWP, no licence is required to operate a scissor lift, irrespective of the height it is capable of reaching. That is, the 11 m limit imposed on EWPs generally, and above which a licence must be held by the operator, does not currently apply to these units, and there is no change envisaged to this ruling in the immediate future.

Note: Though no licence is required, certificated training on scissor lifts is! Experience is also important to safe usage (see 'How to' box on pages 128–129).

Vertical lifts

Vertical lifts (Figures 3.6 and 3.7) serve a similar purpose to scissor lifts (taking people and materials straight up), but are constructed very differently. This variation allows for a lighter construction, as well as offering alternative modes of use such as the fly jib shown in Figure 3.6. Unlike trailer-mounted EWPs or scissor lifts, vertical lifts use neither booms nor scissor-like arms. Instead, they operate in a manner not dissimilar to forklifts – that is, they have a series of frames that are lifted vertically by either a chain or hydraulic mechanism, each section providing lateral support to the one above it.

This form of EWP can be trailer mounted (and so towed into position, much like the trailer-mounted units dealt with previously), made light enough to be manually wheeled around, or may be self-propelled.

Source: Ahern Australia

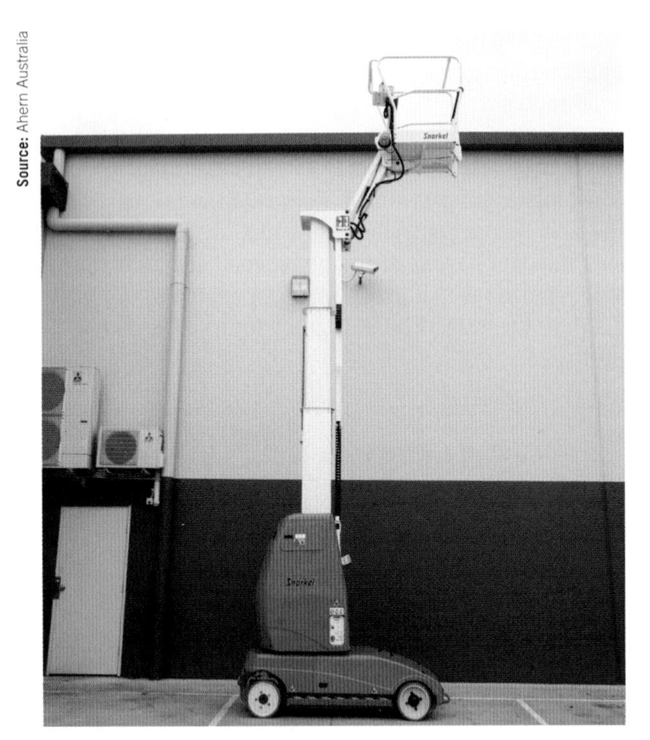

FIGURE 3.6 Self-propelled vertical lift with fly jib

Source: Ahern Australia

FIGURE 3.7 Manually manoeuvred vertical lift with outriggers

Vertical lifts are generally fitted with stabilisers, to be used in much the same manner as those fitted to scissor lifts. This means they are for stability only, as vertical lifts are generally not designed for anything but solid, flat ground or concrete slabs.

As with scissor lifts, no licence is required for this type of EWP, despite some units having the capacity to reach heights in excess of 20 m.

Truck-mounted boom lifts

Truck-mounted boom lifts are technically outside the scope of this book; however, it is worth covering this form of EWP purely for the sake of identification and 'rounding out' your knowledge of EWPs generally.

Truck-mounted EWPs (Figure 3.8) are just that: virtually any of the previously mentioned EWPs mounted on the back of a truck. By using the engine of the truck itself for power (by means of a **power take-off**, or PTO, point), much larger and stronger EWPs may be created. In addition, they may be driven over rough or uneven ground and then, using outriggers, levelled into position. (*Note:* care must be taken when using outriggers on rough or unstable ground. See 'Outriggers, road plates and pigsty packing'.) However, once it is in position and the EWP is raised, the truck cannot be mobilised. So, like trailer-mounted units, they become a static base from which the EWP operates.

Source: Ahern Australia

FIGURE 3.8 A small truck-mounted boom lift

This form of EWP is commonly used for large construction works, but really comes into its own in the erection, service and maintenance of power lines (as there may be a significant distance between one point of service and another). For this reason, truck-mounted EWPs have insulated booms to ensure electricity will not travel through them to the ground (which would endanger the lives of the operators). Both booms and trucks must be tested for electrical isolation regularly, as well as follow a strict maintenance schedule.

Alternative EWPs

In addition to the five groups mentioned above, some EWPs are built for a special purpose and may cross groupings. Such machines are more commonly found in non-construction fields, such as farming (small self-propelled 'cherry pickers') and the airline industry (large truck-mounted scissor lifts and self-propelled booms with custom knuckle arrays). No matter how much experience you have had with EWPs, if you need to operate a unit that you are not familiar with, you must ensure you receive proper instruction on that specific machine, as well as read and comprehend the manufacturer's operating manual.

Modified EWPs

On occasion, you may come across an EWP with which you are familiar, but which has been modified, or you may require to have the unit modified yourself for a particular purpose. In such cases, note that 'modifying' includes *any* drilling, welding, bolting, cutting or screwing – that is, anything that adds to, takes from or otherwise alters any part of the EWP.

To modify an EWP, you *must always*:

- apply for and obtain written approval for the specified modifications through your state WHS/OHS authority
- ensure that the modifications comply with national and state/territory regulations
- ensure they are in accord with all relevant Australian Standards.

Evidence of approval and compliance must be appended to the service and operating manuals and be kept with the machine at all times.

As the operator, you should check the documentation if you suspect the machine has been modified. If you suspect that a machine has been modified and no documented approval is available, you *must not* use the machine.

Prohibited modifications

Some elements of an EWP *must not* be changed. This is because such changes could result in failure or overbalancing of the machine, endangering the lives of operators and/or other workers, or bystanders who may be in the vicinity.

There are two main modifications that must never be attempted:

1. *modifications to the controls*: control units on EWPs are very complex modules that have many inbuilt safety and limiting features. Making changes to the controls or module housing could interrupt these features, leading to the machine being able to overextend and so overbalance
2. *addition of carry baskets or tool bins to the platform*: this form of modification can easily overload an EWP and, again, lead to overbalancing. Carry baskets hanging off the platform railing mean that the weight is not where the designers originally planned in their engineering. Loading an EWP this way can lead to it toppling forwards or sideways. Tool bins can be loaded with an unknown amount of weight, which will be unsighted by the operator and perhaps not be taken into account when determining a safe working load (SWL). Again, this can lead to toppling.

EWP design and compliance

AS/NZS 1418.10 Cranes, hoists and winches Part 10: Mobile elevating work platforms is the Australian Standard to which you should look when considering the purchase or modification of an EWP. This standard covers every aspect of mobile EWPs (referred to as MEWPs), including their design, manufacture, maintenance and, to some degree, use. A further standard applies specifically to the safe use of EWPs: AS 2550.10 Cranes, hoists and winches – Safe use – Mobile elevating work platforms.

Duplication of controls

Within the standard there is one passage that is important for all operators to remember. This passage demands that all mobile EWPs must be equipped with duplicate controls. These controls must be located in a protected position and be accessible from the support surface. In addition, they need to be capable of overriding the controls on the platform so that the platform can be recovered in the case of illness or injury of the operator. It is important that only one control module is capable of operating at any one time.

As the operator of an EWP, you must never work in a machine that does not have functional controls both on the platform and at ground level. AS 2550.10, however, makes further stipulations as to the use of these controls at any given time. Only the platform controls may be used when the machine is being raised, slewed or otherwise positioned with workers on board. That is, only you (the operator on the platform) may raise or otherwise position the machine. This is because a person on the ground cannot see obstructions, and hence potential crushing points, from down below; they are visible only to the operator. The design stipulation is that the base control module should be used to override the platform controls in an emergency situation only.

AS 1418.10 also stipulates that there must be a means of retrieving the platform in the event of power failure. Generally, this is by way of bleed valves in the hydraulic lines, which will be discussed later in the chapter, though the provision of alternative power supplies is allowed for.

Compliance plates

EWPs, like cars, trucks and trailers, will have an identification or compliance plate – generally, on the chassis, drawbar (if a trailer type) or inside the engine compartment. Without this plate the machine is illegal and must not be used.

The plate should show the following information as a minimum:

- *date of manufacture:* the year (the month may be specified, but is not necessary) the machine was originally made. It may also specify the model year.

Like cars, this means it may be built in late 2015, but be a model year of 2016

- *model name and number:* the manufacturer's identifier for that specific machine
- *weight or mass of machine:* refers to the unladen (unloaded) weight of the machine itself. Also known as the **tare weight**
- *SWL:* the load that the machine may safely carry, including occupants. Specified as a mass in kilograms, it may include a statement of the maximum number of occupants. This information (maximum occupants and SWL) must also be clearly marked on the platform, as well as in writing in the logbook and operating manual
- **maximum operating angle**: sometimes called the 'chassis inclination' or 'operating incline', this is a slightly confusing name for what is actually the maximum incline of the ground beneath the EWP. When operating, the machine must of course be made level (by means of the outriggers). Specified in degrees or as a percentage of gradient, there may be two angles given, one for each axis (front to back, and side to side)
- *maximum* **side force**: the maximum force you may push or pull on the side of the machine when elevated (or partly elevated)
- *maximum-rated wind speed* (**wind rating**)*:* the highest wind speed for safe use of the machine. Given as either metres per second or km/h, this is a tricky piece of information given that you will generally have no means of actually measuring the velocity of the wind. It is indicative, however, and tells you at least that you should not be operating the machine in high winds
- **gradability**: specified in degrees or as a percentage of gradient, this is the maximum angle of the ground over which the machine may be mobilised with the EWP in the lowered position
- *statement of compliance:* a statement that the machine complies with the relevant Australian Standard (AS 1418.10).

Other information

The following information may be available on the plate, or should be available in the operating and/or service manuals:

- *in service date:* the date (day/month/year) that the machine was 'commissioned' or first brought into service. As with cars, this may be several months or even a year or more after manufacture. All service schedules will be based on this date, so it must be entered into the front page of the logbook
- *maximum* **platform height**: the height the floor of the platform will be above the supporting floor or ground when the machine is at maximum extension. Measured as a vertical distance
- *maximum reach:* the distance the EWP can reach horizontally from the central axis of the turntable
- *stowed height and width:* the overall dimensions of the machine when in the stowed or travelling position.

KEY POINTS

All EWPs must:

- have compliance plates affixed in a visible place on the frame
- be set up *level and stable* prior to operation
- have duplicate controls
- have outriggers, or if self-propelled without outriggers, have solid tyres.

MODIFICATIONS

You must apply through your relevant state WHS/OHS authority *before* making *any* modifications.

A modification is *any* drilling, welding, bolting, cutting or screwing on *any* part of the EWP.

LEARNING TASK 3.1

1 **Circle 'True' or 'False'.**
Modifying an EWP refers only to changes to the controls, the main boom or scissor lift components, or anything to do with the engine or hydraulic system.
True False

2 **If you need to modify an EWP, you must:**
a apply for and obtain written approval for the specified modifications through your state WHS/OHS authority
b ensure that the modifications comply with national and state regulations
c ensure they are in accord with all relevant Australian Standards
d all of the above

COMPLETE WORKSHEET 1

Logbooks and manuals

All EWPs will come with a logbook and operating manual upon purchase. (This is mandatory.) The purpose of these texts is to provide important service and operational information.

The logbook

By law, the logbook must remain with the machine at all times. Generally, you will find it stowed in a yellow waterproof pouch permanently attached to the platform (see Figure 3.9). Sometimes it may be in a black waterproof box, also on the platform (see Figure 3.10).

FIGURE 3.9 Soft case housing for logbook attached to EWP platform

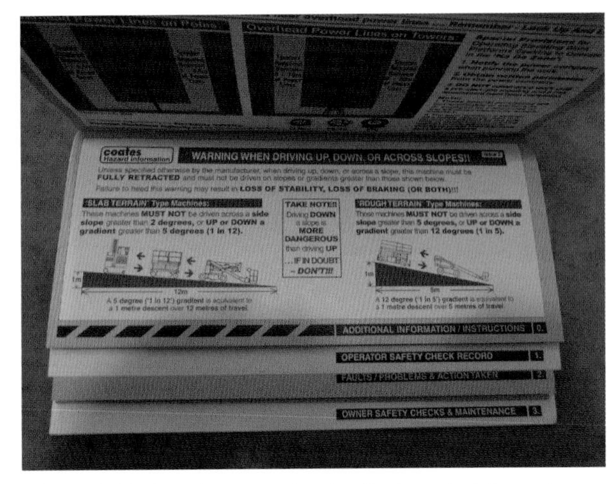

FIGURE 3.11 Open logbook showing the layout of the various parts that need to be addressed by the operator prior to and after use

FIGURE 3.10 Common hard-case logbook enclosure on an EWP

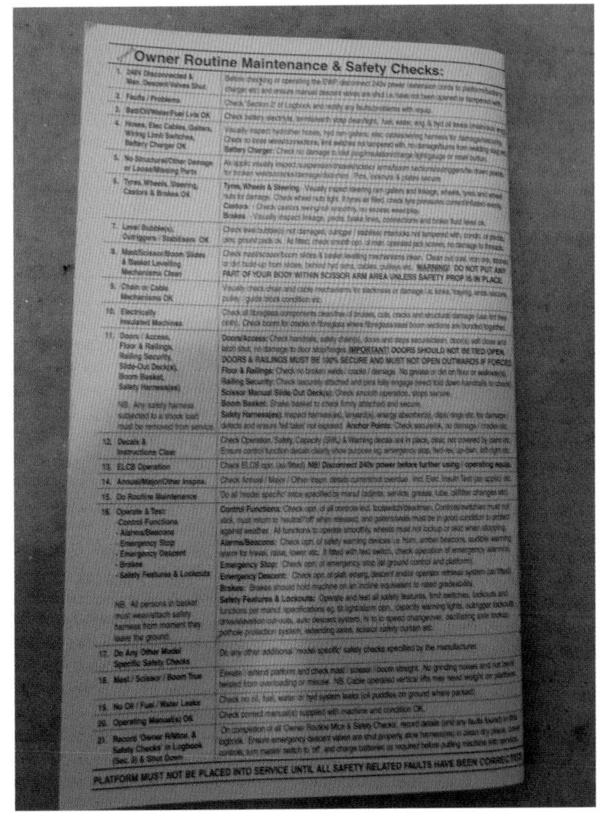
Owner Routine Maintenance & Safety Checks:

PLATFORM MUST NOT BE PLACED INTO SERVICE UNTIL ALL SAFETY RELATED FAULTS HAVE BEEN CORRECTED

FIGURE 3.12 Back cover of EWP manual giving routine pre- and post-operational safety checklist

An EWP without a current logbook is not an EWP – it is a questionable piece of machinery that cannot be used.

The logbook provides the operator and other inspectors with the following:

- information identifying the machine (serial number, make and model)
- structured pre- and post-operational checklists
- maintenance schedules and requirements
- a means of documenting (logging) that the above have been conducted
- a means of documenting (logging) any repairs required or completed
- a record or history of use and users (who and when).

Most logbooks are divided into three parts and a preface (see Figure 3.11), where:

- the preface provides information about the machine itself (serial number, make, etc.)
- Part 1 is the record of use
- Part 2 is the faults reporting section
- Part 3 is the maintenance record.

Generally, on the front and back covers (or inside these) you will find the pre- and post-operational checklists (see Figure 3.12). We will discuss the application of these checklists in the section 'The safe use of EWPs: Part 2 – checks, loads and positioning' later in the chapter.

If you are new to the machine, always check that the serial number listed in the preface of the logbook matches the number on the compliance plate of the machine itself.

Before using the machine, you must fill out Part 1 of the logbook, including:

- the date and time you begin to use the machine
- your (operator's) name

- results of the pre-operational safety check
- your (operator's) signature
- your (operator's) card number or licence number (where the machine requires a licence for use)
- supervisor's name/signature, if applicable.

Before using the machine, you must check Parts 2 and 3 of the logbook for the following:

- reported faults have been signed off as repaired by a technician
- services have been conducted as required* (not more than three months apart).

If either the services or the faults have not been completed or repaired, then you *must not* use the machine. You must report the situation to your supervisor, who must have the machine either repaired or replaced.

Fault reporting

If you find a fault during the pre-operational safety check (or during use), you must report the fault immediately to your supervisor/manager and then enter the following information in the faults section of the logbook (Part 2):

- your name
- details of the fault
- date and time of finding and reporting the fault
- name and contact number of the person you reported the fault to.

Operating manuals

Logic suggests that at least a copy of the operating manual should be in the same place as the logbook – that is, with the machine (see Figure 3.13). (Sometimes it is kept in a separate box or locker, either on the platform or near – and on occasion in – the engine compartment.) Sadly, this is not always the case. If you are unfamiliar with the machine, you must contact the owner or your supervisor for familiarisation training and a copy of the manual.

The manual generally provides you with the following important information:

- a history of any revisions to the publication
- safety information
- user responsibilities
- preparation and inspection checklists and details
- stowing and mobilisation information (towing or driving, depending upon EWP type)
- overview of the controls
- overview of safe operation
- emergency procedures
- maintenance schedules and general service details
- specifications and capacities
- schematic diagrams showing and labelling important parts and their location.

* *Note:* Service schedules vary from machine to machine, and will also change with the frequency of use. Always check the operator's/service manual for required services, which will generally be stated in hours of operation. Irrespective of the hours of use, services must be conducted not more than three months apart.

FIGURE 3.13 The preferred location of an operating manual, in the case or enclosure on the EWP platform

Note: Some manuals may also have an inspection and fault repair log included. However, it is general practice that this information will only be entered into the mandatory separate logbook discussed earlier. The operating manual should also provide information on the type of power source that should be connected to the EWP and how this is protected.

KEY POINTS

- Logbooks must remain with the EWP at all times; an EWP without a logbook is not a licensed EWP and *must not be used* under any circumstances.
- Fill out and check logbook service, fault and repair history *before* using the EWP.

LEARNING TASK 3.2

1 **Circle 'True' or 'False'.**
You can use an EWP that is not equipped with a logbook.
True False

2 **Circle 'True' or 'False'.**
You can use an EWP that is not equipped with a user's manual.
True False

Safety around hydraulics

Elevating work platforms are, almost universally, operated by way of hydraulics. Hydraulic systems use confined, pressurised fluids passing through braided hoses and couplings (see Figure 3.14) to increase or otherwise modify an imposed force. In EWPs, they are used to drive pistons, which in turn move major components such as booms, jibs, scissor arms and the like. The fluid used in most hydraulic systems is a petroleum-based oil of a clear, light yellow colour with a very faint smell common to most lubricating oils.

Source: Shutterstock.com/Paul_K

FIGURE 3.14 Hydraulic tubes, fittings and levers

Unfortunately, the risks associated with hydraulic systems are often poorly understood by most operators of equipment using this form of power transfer. Indeed, most training on equipment that uses hydraulics omits any reference to the associated risks at all.

The following is a basic breakdown of the things you should keep in mind when working around hydraulic systems.

Hydraulic oil

Note that:

- hydraulic oil is toxic to humans and animals. It must not be ingested (swallowed), allowed to enter the bloodstream, or its fumes or vapours inhaled
- it is highly viscous. Even at room temperature, hydraulic oil is easily absorbed and may enter the bloodstream through the skin without pressure
- like many thin or viscous oils, it is difficult to see in bright sunlight
- at operating temperature (around 150 °C), the risk of vapour inhalation is greatly increased. The low odour further increases the risk of overexposure
- hydraulic oil has a flash point (spontaneous combustion when mixed with air) of between 150 °C and 300 °C. This means leaks from a system in operation can catch fire very easily. Due to this it's important that you note the location of fire fighting equipment whenever working with EWPs.

Hydraulic systems and hoses

Note that:

- hydraulic systems and hoses contain hydraulic oil with an operating temperature approaching 150 °C
- small puncture leaks can emit oil at 3000 psi, which can penetrate the skin as easily as a syringe, or cut like a scalpel
- small puncture leaks can become a fine, powerful mist that can travel as far as 10 m from the hose itself. A heat source that might ignite this stream is therefore often easily in range (such as heaters, welding equipment, cigarettes or hot metals)
- hydraulic hoses are constructed of braided metal encased in rubber (see Figure 3.14). A bulging hose may indicate a failure of this braiding and the hose must be replaced immediately even if the exterior rubber of the hose appears perfectly sound (undamaged).

Risks from hydraulic systems

Injection wounds and burns are the greatest risks from hydraulic systems. With injection wounds, often the penetration to the skin is quite small, but the burning fluid then effectively replaces the blood and, following the veins, causes massive burns to the surrounding tissue and bones. Such wounds frequently lead to amputation of limbs.

The risks from hydraulic systems can be summarised as follows:

- injection injuries (fluid forced into the body and bloodstream by a fine, high-pressure spray – injection to the upper body can penetrate and destroy muscles and organs)
- burns (usually third-degree burns, as the oil will penetrate very deeply, very quickly)
- crushing injuries (hoses snapping off fittings at high pressure will whip and flail around very fast and hard, breaking bones and tearing tissue).

HOW TO

PRECAUTIONS WHEN WORKING WITH, OR INSPECTING, EQUIPMENT USING HYDRAULIC SYSTEMS

As part of the pre-operational checks for EWPs, you must inspect hydraulic hoses, lines and pistons. You also must check the hydraulic fluid levels. When conducting these inspections, be sure to do the following:

- Do not let the oil come in contact with your skin. If it does, wash immediately with water-free hand cleaner. (This will reduce absorption.)
- Wear chemical-resistant gloves and safety glasses when topping up hydraulic fluids (much as you would when topping up battery fluid).
- If clothing becomes contaminated, remove it, and wash your skin thoroughly. Clothes must be washed completely before reuse or be discarded.
- Never look for a pinhole leak in a hose or coupling by running your hands or fingers over the lines. While wearing chemical-resistant gloves, use a piece of heavy cardboard or light wood to detect it.

If a leak is detected:

- shut down the machine immediately (if running)
- make no attempt to adjust or tighten couplings, or change hoses if a leak is detected. Only a fully trained technician should be allowed to do this work
- ensure there are no ignition (hot) sources anywhere near the machine
- if the leak is spraying for some distance from the machine, establish a 'no go' area around the machine. Avoid breathing in the vapours from fine mists of oil
- contact the owner or your supervisor. The machine is not to be operated until the leak is fixed.

In completing the training for the competency unit CPCCCM3001: Operate elevated work platforms up to 11 metres, you are not trained at any level to work upon or maintain hydraulic systems. You may inspect, but not adjust or repair. To exceed this limitation is to risk significant injury or death – to you, and to others.

LEARNING TASK 3.3

1 **Circle 'True' or 'False'.**
Only a qualified hydraulic line technician is allowed to inspect the hydraulic components of an EWP.
True False

2 **When inspecting a hydraulic hose line for leaks you should:**
  a place your hand lightly around the hose and be sure to move you hand down the full length of the line
  b while wearing chemical-resistant gloves, run a piece of heavy cardboard or light timber along the surface to spot the leaks
  c wash any hydraulic fluid from your hands immediately should any get on you accidentally
  d both 'b' and 'c'

COMPLETE WORKSHEET 2

Rescue plan

In construction there is always a risk of something going wrong. If and when things do go wrong, it's important to have measures in place to resolve the issue as safely as possible. When working with an EWP this measure is a 'rescue plan'.

In the event someone falls from an EWP and is left hanging in a IFAS there is a real risk of suspension trauma. Suspension trauma occurs when a person's body is suspended in an upright, vertical position, such as while awaiting rescue after an arrested fall in a harness.

In an event like this, a rescue plan will detail the steps that should be taken to essentially rescue the person from the situation.

Before accessing an EWP, ensure you have reviewed the rescue plan and have a good understanding of what you are required to do if an accident or issue does occur.

The safe use of EWPs: Part 1 – spotters, hazards and controls

Having worked your way through the first part of this chapter, you should by now be familiar with what an EWP is, the various types generally available, and what should come with them (logbooks and manuals, for instance). This section is about actually using one.

The safe use of EWPs is broken into two parts. This first part looks at what may be called 'job planning': selecting the right EWP, and locating it so that it may be used safely. In so doing, it also provides an overview of the controls, what you need to know and do before entering the machine, and the correct use and attachment of an individual fall-arrest system (IFAS). Part 2 (which follows in the next section) then describes the pre- and post-operational checks you must do, positioning the machine and safe stowing for transit.

Before considering any of these topics, however, we will look briefly at a most underrated, yet critical, role with respect to safe use of an EWP: the spotter.

The spotter: a critical role

A spotter is an alert, skilled and knowledgeable person with experience in the work being undertaken. Spotters (also known as 'safety observers') have the following vital responsibilities:

- warning operators and other persons of potential dangers
- stopping the work being undertaken due to a perceived risk
- carrying out recovery procedures when a machine has failed or an operator is ill or injured and cannot lower the platform themselves
- instructing rescue personnel on the operation or nature of the EWP, or any equipment or materials on the platform, in emergency situations.

The role requires a spotter to be constantly on the lookout for the health and well-being of the operator and to ensure that equipment, people (workers, the public, and the equipment operators themselves), and in some instances, animals, do not cross an agreed exclusion or 'no go' zone. (Spotters may also need to advise the operator of potentially dangerous aggressive bird behaviour.) It is a role made all the more difficult because the 'line' not to be crossed is often invisible (see Figure 3.15).

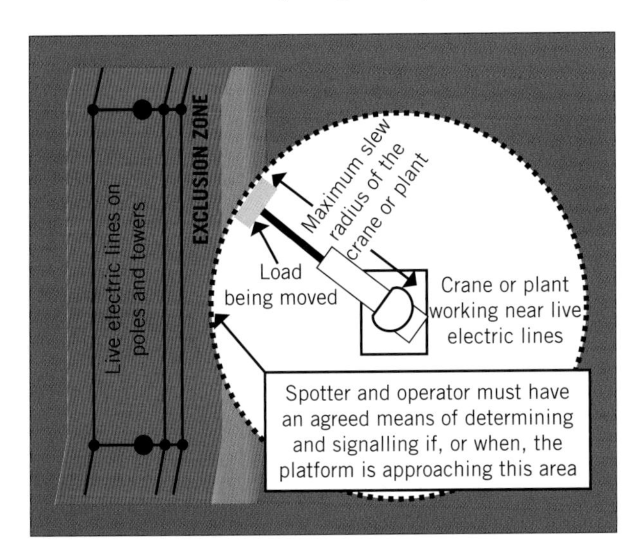

FIGURE 3.15 Example of exclusion zone around electrical lines

Spotters may be required when using EWPs around other workers, in public spaces, near power lines, where overhead or potential crushing obstructions exist, or close to voids, trenches or the like.

When barriers and signage have been established, the role of the spotter is fairly straightforward: ensuring that people stay behind them. When close to power lines, however, it is a matter of being able to judge the distance from a line when looking up from below (though some operators use a spotter on the platform as well as below).

As the spotter is frequently below and often at the extremities of the exclusion or 'no go' area, the role of communication between spotter and operator comes into question. Without clear and concise communication between the two parties, the spotter may as well not be there at all. Communication may be by two-way radio, hand signals, whistles or horns. Whichever approach is chosen, both parties (operators and spotters) must be fully conversant with the meanings of any signals or terms used. These days, the best approach is by means of two-way radios (being cheap and easily obtained), though a backup system is always advisable (preferably whistles or horns) in case of battery or signal failure or when an emergency stop is required.

The role of the spotter is critical to the safety of operators, other workers and the general public. It is not a role to be undertaken lightly, nor mistakenly considered a 'no brainer' and so only for the less skilled. If you are instructed to be a spotter, ensure you fully understand your role, the work being done by the EWP operator, and the capacities and operation of the machine, and are very clear on any clearance distances to be maintained (and from what). The best way to do the latter is to have the operator provide an example of the distance while in discussion with you, so that you have a mutually agreed understanding. If in doubt, ask, and run through it again. Keep your mind on the job – lives depend upon it ... upon you.

Now, let's look at choosing the appropriate EWP for both location and task.

KEY POINTS

Spotters must be:

- knowledgeable and alert
- able to judge distances
- in constant communication with the operator.

Choosing the right EWP

Your choice of EWP is critical to your capacity to complete a task safely, in a timely manner and cost-effectively. In making this decision there is a very broad range of questions you must ask yourself, and at times, others. You must learn to think about 'context'. By context we mean the location (access, space, height of task, etc.), the environment (wind, sun, rain, soil types, the terrain and the like), indeed everything and everyone that can or may have an influence upon the task being considered. The question goes much deeper than simply, 'How will I get up there?'.

In addition, you should work well within your own limits and those of the machine. This means choosing the bigger machine if you would frequently need to use a smaller unit at its full capacity. Otherwise, both you and the machine may be stretched to your limit, which leads to stress, tiredness and, ultimately, poor decision-making. Remember that both people and machines, when constantly at their limits, are more likely to fail; and failure at heights has serious consequences.

The 'On-site' box that follows provides you with just some of the questions you should be considering when making your decision. It is not exclusive (i.e. you may find that in your particular context there are other, more important, questions confronting you), but it does give you a starting point on how you should be thinking about your task.

ON-SITE

CHOOSING AN EWP

The following are a few of the key questions you should get answers to when deciding which EWP is right for your particular task.

- How will you access the site and the specific area of work?
- What surface will the machine be working on (stable, rough, flat, level, etc.)?
- How high does the platform need to be for you to do the work?
- What distance might the platform need to reach for you to do the work?
- Does the machine need to articulate (bend) over obstacles to position the platform?
- How much weight will be on the platform (people, tools, materials)?
- What sort of work will you be doing – might you be putting side pressures on the platform, or doing electrical work, for example?
- Will the EWP need to be mobilised frequently, or can it access all work areas from one or two positions?
- Are there any overhead obstructions and crushing risks?
- Will the machine be functioning in a confined space (creating exhaust fumes)?
- How frequently or for how long will you need the machine?
- Are there any electrical hazards present (e.g. power lines)?
- Are there any underground or subsidence hazards (recent backfilling or nearby trenches)?
- Are there any noise restrictions – such as when working in or around hospitals or the like?

Other issues you may need to consider:

- refuelling needs
- permits and approvals for working in public spaces, environmentally sensitive areas, across or close to boundary lines
- access to alternative power supplies if these will be needed for emergency recovery procedures
- public access
- access for other workers
- scheduling of other works in or around the EWP location(s).

To gain this information, you may need to talk to site supervisors and property owners (and, on occasion, neighbours and councils), read plans and actually conduct a site inspection. Only after you have gathered this data are you in a position to make an informed decision.

KEY POINTS

When choosing an EWP you should:

- think *context* as well as task
- choose a machine that is working *within its operational limits*.

Locating the EWP

Good job planning starts, continues and ends with site inspection: knowing the context of the area where the work will be undertaken. In order to choose the right EWP, you should already have a clear understanding of the task at hand and where it will be done. Your questioning process will have included:

- conducting a site inspection and site supervisor consultation
- holding discussions with other stakeholders
- determining public and/or worker access
- determining and gaining any required permits
- identifying hazards.

In locating the correct EWP, we need to look at these areas a little more closely, then consolidate the information obtained into a job safety analysis (JSA) and/or safe work method statement (SWMS). Towards the end of this process, we will look at one hazard – power lines – in detail, before determining the best place to put the EWP.

Site inspection and consultation

Through this process, you will determine the layout of the site, your access and egress routes, and other factors discussed earlier, to determine the right EWP for the job. In addition, you will need to find out from the supervisor the following:

- site policies and procedures (this may be as simple as signing in and out, for example)
- site induction procedures
- the job schedule for other trades
- support structures for surrounding works (including underneath the slab, if you are working over a basement – you may need an engineer's assessment)
- particular site hazards (trenches, backfill, subfloor works, wind or other exposed condition issues)
- public and worker access.

Having consulted with the site supervisor, you will now have identified the other trades with whom you must also discuss your planned work.

ON-SITE

A NOTE ON EXCLUSION ZONES AND 'NO GO' AREAS

It is important to note the difference between an exclusion zone and a 'no go' area as they are applied in practice.

- A 'no go' area is one into which only the EWP, its operator and those immediately concerned with the work requirements being undertaken using that machine (such as a spotter) may enter.

>>

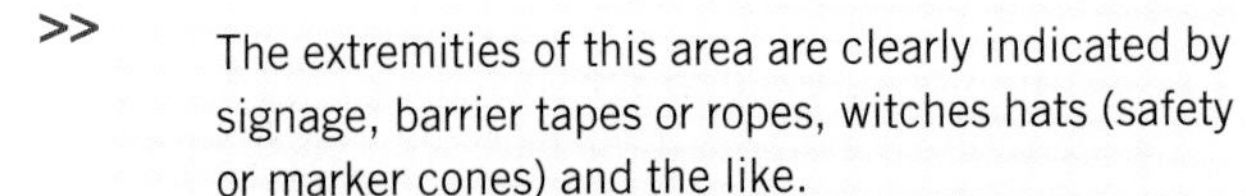

The extremities of this area are clearly indicated by signage, barrier tapes or ropes, witches hats (safety or marker cones) and the like.
- An exclusion zone is an area that is dangerous for the operator to allow the EWP to enter. A common example is set distances from electrical supply lines, as outlined under 'Locating the EWP'.

Discussions with other stakeholders

What you need to determine here is how your work is going to influence or interrupt the work or general interests of others. People who may have a 'stake' (whose lives or interests may be influenced by your activities) in work that is about to happen may include more than just other workers. Other stakeholders may include:

- the local council
- owners and workers in other businesses
- pedestrians
- owners of neighbouring properties
- road and transport authorities
- police
- service suppliers (telecommunications, power, water, gas, etc.).

Not all of these are going to be able to have personal input, but all may need to be notified of interruptions, and some may have the right to limit or even stop your intended works. It is better to have your discussions with them prior to proceeding, rather than have expensive machines sitting idle while you sort it out mid-job.

With regard to other workers on the site, you must look carefully at where they are going to be and ensure that either:

- they stop working in those areas that you will need to be above (in slewing or otherwise positioning the EWP)
- you position the machine so that it does not need to pass above these areas.

There may be other considerations you should discuss with them, such as:

- exhaust fumes
- noise
- interruption to power supplies
- limited or blocked access (for materials and/or workers)
- additional personal protective equipment (PPE), such as hard hats
- liaison with a trained spotter from your work team.

On some sites, these discussions may need to be with collective worker organisations such as unions. Again, it is better to do this before, rather than after, you have set up and begun working. Stopping mid-job is expensive.

Public and/or worker access

With the above information, you can then determine the best approach to public and/or worker access and egress around your work area. Note that this is always 'around' your work area, not through it, while you are operating the EWP.

The best solution to 'crowd control' (be it other workers or the public) is the establishment of a 'no go' area or zone. Means of establishing 'no go' areas include:

- witches hats
- tape and mesh barriers
- non-climbable fencing
- warning signs
- barricades and barriers
- traffic control personnel
- trained spotters.

Part of your discussions may include a 'down time' for your machine. In this way workers, including perhaps pedestrians from other places of work, may pass through the 'no go' area while the machine is lowered and not in operation. This may be for lunch periods, or specific times when the public, traffic, other workers or materials have a particular need to get through.

Determining and gaining required permits

While establishing a 'no go' area may be the best solution to keeping people and traffic out, you may need to gain a permit to be there yourself. Indeed, you may need to gain several permits. Authorities to which you may need to apply include many of the stakeholders mentioned previously:

- road and traffic authorities
- police
- local council
- electricity authority
- water board
- gas company
- telecommunications companies.

To determine to whom you should apply, the best approach is to assume that you will need a permit from one or all of the previously mentioned authorities if you:

- are near any of the above services
- block traffic or public access
- operate (even in slewing) over, or close to, property other than the work site.

In other words, contact *them* before they contact *you*. They may then contact other stakeholders you have not thought of, paving the way for better relations all round.

Hazard identification

It is mandatory at all places of work that you carry out a risk assessment prior to locating and/or operating an EWP. As discussed in previous chapters, this may be by means of a SWMS or a JSA, or a combination of these.

Due to the research and discussions you have just concluded, this process is now fairly straightforward, being effectively a summary coupled with an action plan. Hazards you may have identified may be classified under the following headings:

- sub-operating level:
  - backfill
  - excavations
  - basements
  - lower floor-level supports

- underground services (gas, water, electricity, etc.)
  - voids, staircases, access holes and trenches
- operating level:
  - angled surfaces
  - rough or uneven terrain
  - unstable ground
  - ramps
  - obstructions
  - poor light
  - traffic (vehicular and pedestrian)
  - other tradespeople
  - wet or slippery surfaces
  - confined spaces (fumes)
  - dangerous materials storage
  - power supplies (extension leads and other low-level supply lines such as gas or fuel)
- overhead:
  - electrical supply lines (power lines)
  - climatic conditions (wind, rain, etc.)
  - maximum headroom
  - crushing obstacles (pipes, obtrusions, other plant and equipment)
  - other works being undertaken at a higher level
  - nesting birds (plovers and magpies, in particular).

Power lines

Power lines impose very particular hazards due to electricity's penchant for bridging or 'jumping' to the nearest earth-connected source. An uninsulated EWP plays the earth role all too well. Due to this, AS 2550.10 specifies:

- the distance you must be from various line voltages
- when a spotter must be used.

In consultation with the relevant power authority, further stipulations may be made, such as:

- the type of EWP you might use in particular situations (e.g. insulated)
- when insulating covers must be applied to lines, and by whom (often called 'tiger tails' due to their visual black and yellow striping).

While AS 2550.10 provides a national framework, slightly different clearance distances have been set by each state and territory WHS/OHS authority. These differences can complicate matters, as they sometimes allow you to get closer than the Australian Standard stipulates. In light of these differences, it is best practice to follow the recommendations of the Elevating Work Platform Association of Australia (EWPA) – that is, whenever any part of the operator or the EWP is likely to breach the minimum distance of 6.4 m from distribution lines (standard power poles and the like) or 10 m from transmission lines (lines on towers), you must apply for written permission from the relevant energy authority (EWPA, 2013).

Visit the EWPA website at https://ewpa.com.au/resources/information-sheets > EWP Safety Guidelines & Information Sheets > Distance from Power Lines for a full listing of the current state and territory guidelines on power line clearances.

AS 2550.10 requires:

- for distribution lines up to and including 133 kV (standard power poles and the like), you must go no closer than 6.4 m, or 3 m with a qualified spotter
- for transmission lines greater than 133 kV (generally, large poles and towers), you must go no closer than 10 m, or 8 m with a qualified spotter (EWPA, 2013).

These distances are shown graphically in Figures 3.16 and 3.17.

In these two figures, you can see that you must never cross or work 'above' a power line, no matter what the distance. This practice is dangerous because the boom beneath you may bridge the safe distance zone without you being aware of it.

Consolidation: creating a SWMS

You are now in a position to consolidate this information into a SWMS document, as outlined in Chapters 1 and 2. In summary, the approach is as follows:

1. *Identify* the hazard or potential hazard.
2. *Assess* the level of risk and determine the priority, using a risk management matrix. (See Appendix 1 in Chapter 2, Table A.2, for an example risk management matrix.)
3. *Control* the risk using the hierarchy of control shown in Figure 3.18.
4. *Review and evaluate* the risk and the control measures on a continual basis.

In creating your SWMS, be sure to include input from as many of the relevant stakeholders as possible. All workers involved in the task must agree to and sign the finished document.

Having documented an agreed process, follow it! If, upon reflection, or in the doing of the task, it becomes obvious that the process needs changing, do so – but again, all those involved need to be advised of the changes and sign off on them.

Determining the best spot for the EWP

It is during completion of this last step (the SWMS or JSA) that you will have determined the best location(s) for your EWP. You will have looked for a location that, where possible, eliminates the risks you have identified, or failing that, has minimal reliance upon spotters and traffic control personnel (which is subject to human errors).

While all this may seem like an arduous process at first reading, an experienced operator or work team can complete it in a very short time frame. The 'art' is in knowing what to look for and, with sufficient experience, what your options are in any given context. While you are building up that experience, listen to what others have to offer, particularly more experienced operators, site supervisors and workers.

Having now decided where to locate the EWP, it is time to look at the various controls and what they do, after which we can move on to our pre-operational checks.

Source: Based on EWPA Training Resources (2013), p. 2.

Standard distribution lines (133 kV) on poles
NO GO ZONE
For the platform or the person located in the platform
Spotter required between 3–6.4 m of power lines
Anywhere above power lines and within 3 m each side and 3 m from the bottom
Spotter required between 3–6.4 m of power lines
Open area outside 6.4 m of power lines
Open area outside 6.4 m of power lines
3 m
3 m
3 m
Note: Always apply to the relevant energy company if there is a risk that you or the EWP might enter the yellow area.

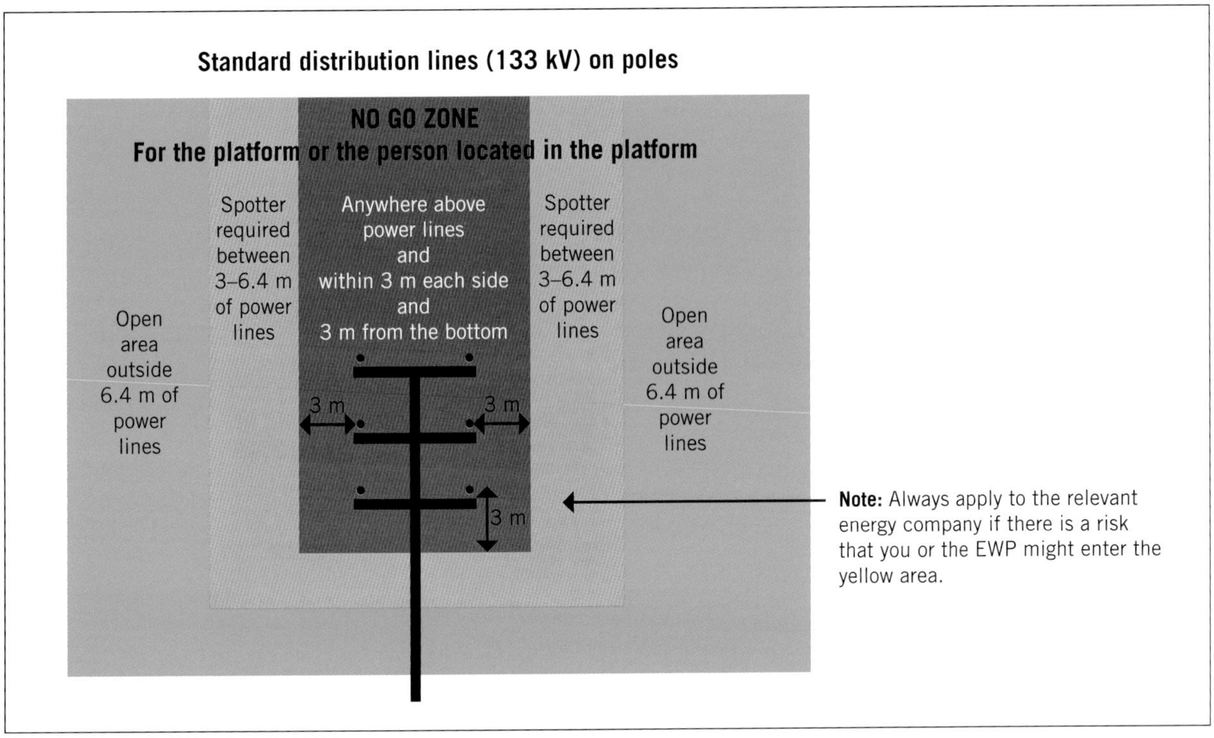

FIGURE 3.16 Standard distribution lines (133 kV) on poles

Source: Based on EWPA Training Resources (2013), p. 2.

Transmission lines (greater than 133 kV) on towers
NO GO ZONE
For the platform or the person located in the platform
Spotter required between 8–10 m of power lines
Anywhere above power lines and within 8 m each side and 8 m from the bottom
Spotter required between 8–10 m of power lines
Open area outside 10 m of power lines
Open area outside 10 m of power lines
8 m
8 m
8 m
Note: Always apply to the relevant energy company if there is a risk that you or the EWP might enter the yellow area.

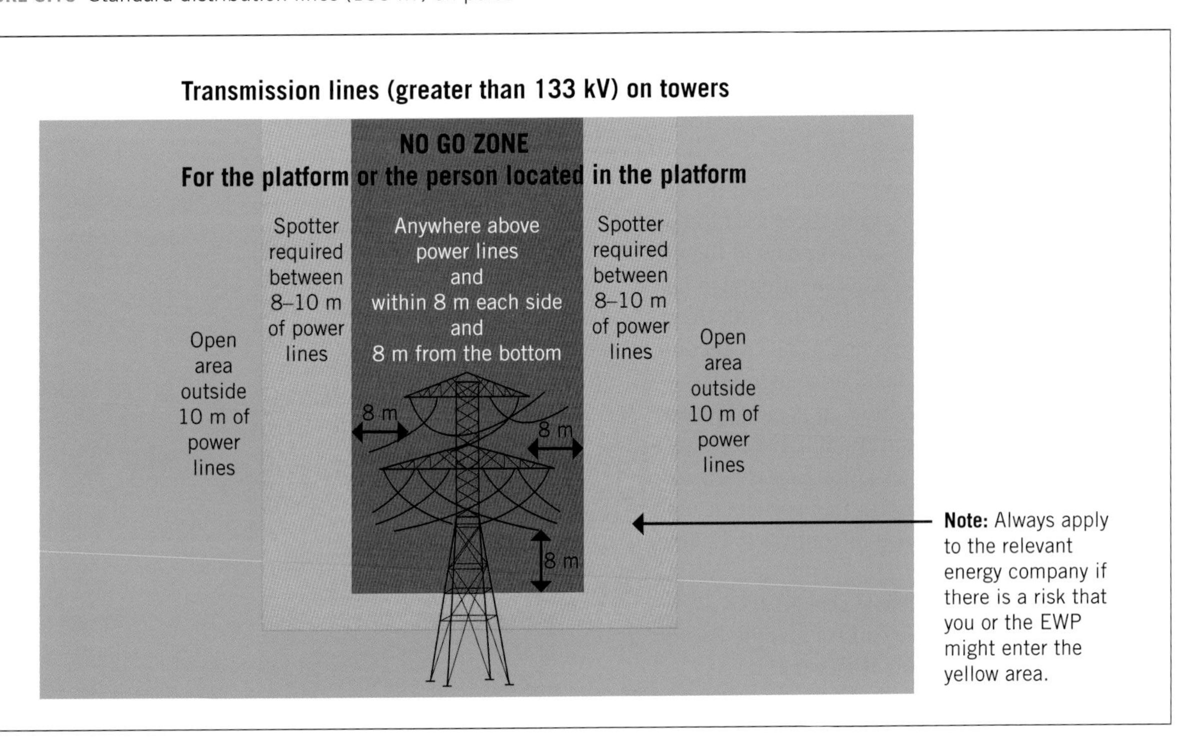

FIGURE 3.17 Transmission lines (greater than 133 kV) on towers

KEY POINTS

When locating an EWP you should:
- conduct a site inspection and discuss the task with the site supervisor
- discuss the task with other stakeholders
- identify public and/or worker access requirements
- determine if any permits are required and, if so, obtain them
- identify hazards and create a SWMS or equivalent document
- the ultimate location of the EWP should reduce exposure to identified hazards and reduce risks, while minimising the need for spotters and traffic control personnel.

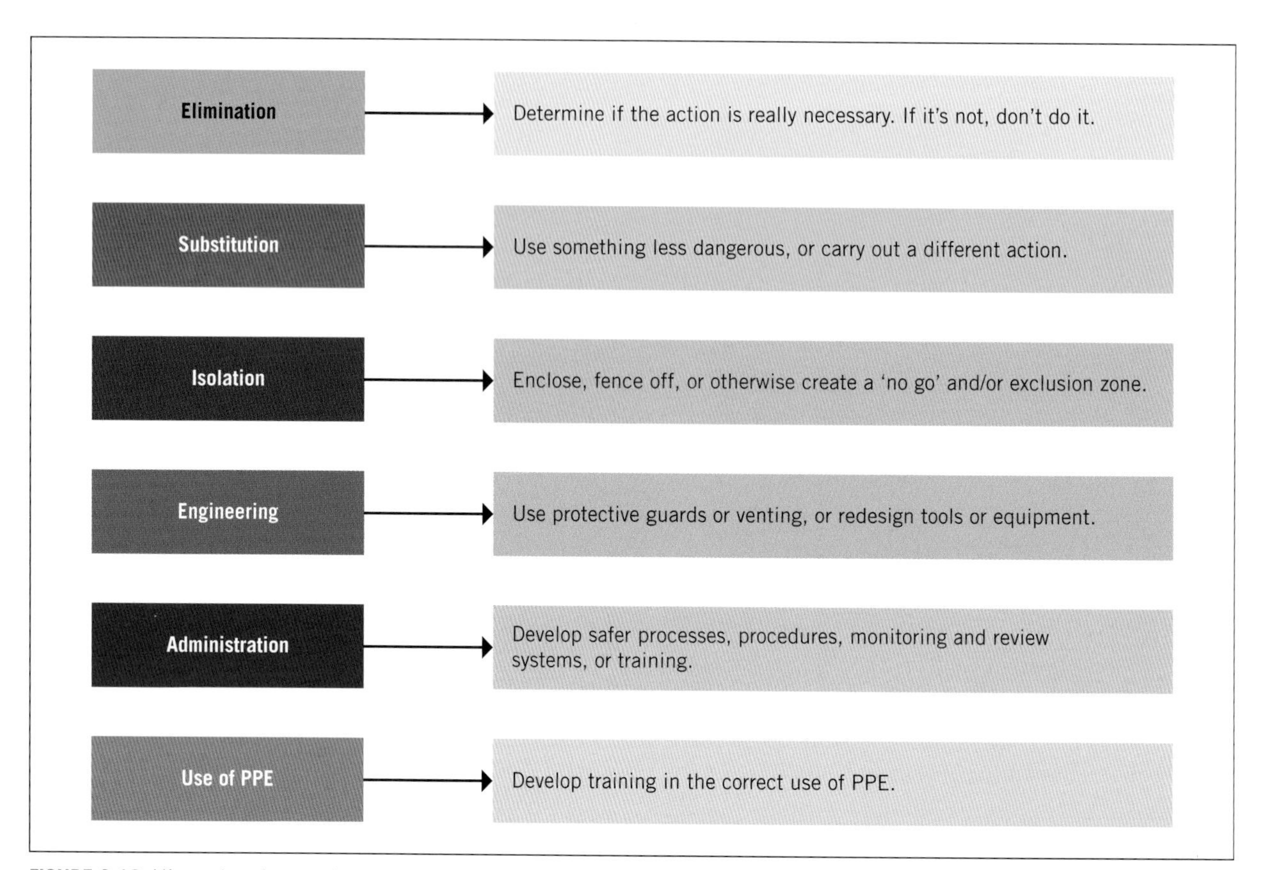

FIGURE 3.18 Hierarchy of control

The control modules

Unfortunately, there are more variations in the control modules available on EWPs than there are types of EWPs. Given this situation, an overview of the most common controls is all that shall be attempted here. As always, you will need to be familiar with the controls of each specific machine you use. While sometimes an experienced operator can familiarise themselves with the aid of the operator's manual, it is recommended that you are always 'checked out' (familiarised with the controls and actions) in a machine that is new to you by someone with experience with that particular unit.

There are three basic sets of controls to be found on most EWPs:

- outrigger control module (Figure 3.19)
- base, ground or chassis control module (Figure 3.20)
- platform control module (Figure 3.21).

A fourth set, remote controls, is sometimes provided for particular machines. These may be fully remote (working off radio signals) or, more generally, be operated by a cable attached securely to the machine.

To these modules may be added the key switching, safety or recovery controls, such as:

- deadman pedal (Figure 3.22)
- base control/platform control override switch (as shown in Figure 3.20)
- hydraulic bleed valves (recovery valves)
- engine ignition switch (as shown in Figure 3.20)

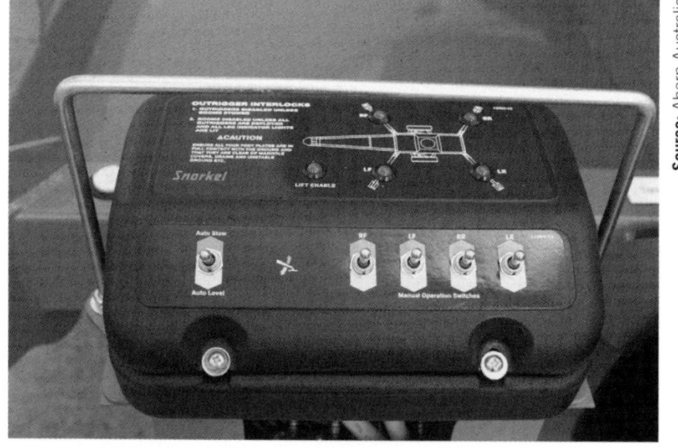

Source: Ahern Australia

FIGURE 3.19 Outrigger control module

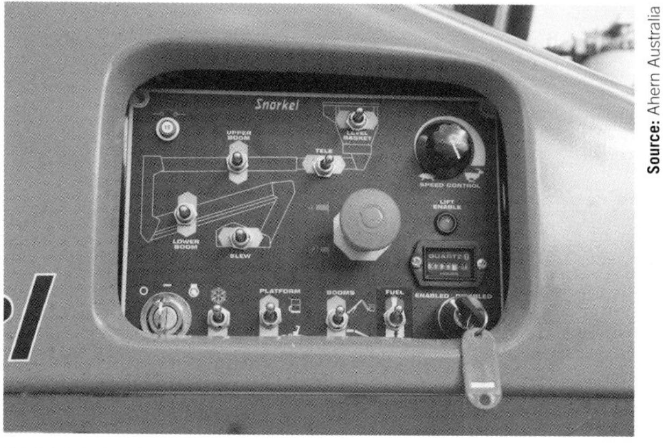

Source: Ahern Australia

FIGURE 3.20 Base control module

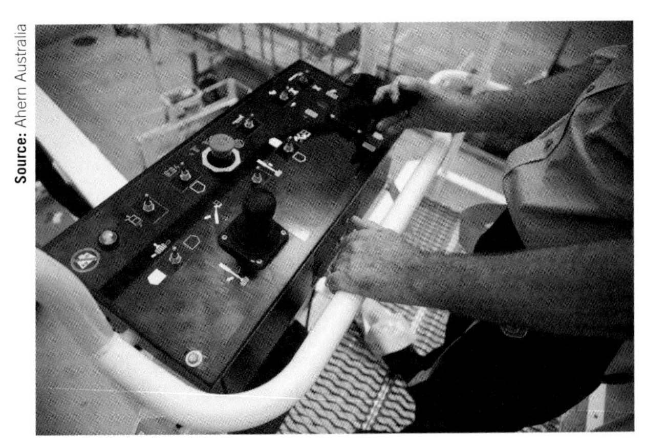
Source: Ahern Australia

FIGURE 3.21 Platform control module

Source: Ahern Australia

FIGURE 3.22 Deadman pedal

- kill switch (as shown in Figure 3.20)
- fuel switch (as shown in Figure 3.20).

Apart from the occasional vertical lift (which may work with chains), EWPs are operated almost exclusively by means of hydraulics. This is because they offer the most powerful and reliable way of raising or lowering heavy masses. (Hydraulics do, however, carry some inherent risks, which were discussed earlier in the chapter: see 'Safety around hydraulics'.)

The more usual control configuration of smaller EWPs (those operating under 11 m) are effectively valves that allow this hydraulic fluid into or out of pistons. These pistons or cylinders are attached by pivot points to the various booms and stabilisers. As the pistons extend or retract, they push or pull the relevant components into the position you require. Other configurations may use electric servo switches to open the hydraulic valves remotely.

Control functions

There are a variety of controls on any given EWP depending upon the type. Generally, however, the controls fall into one of three categories: outrigger controls, base controls and platform controls.

Outrigger controls

The stabiliser or outrigger controls (Figure 3.19), generally mounted on the chassis or base, are in this case simple levers operating hydraulic valves – four levers, four stabilisers, each giving up-and-down actions on the one lever. Near the outrigger controls will be some form of level-indicating device – generally, a blister or bubble level (Figure 3.23) or two small spirit levels set at 90 degrees to each other. In some cases, levelling is electronically determined and a light will come on to indicate the correct position.

FIGURE 3.23 A level-indicating device may be in the form of a blister or bubble

Base controls

Figures 3.20 and 3.24 show a simple valve system of levers (common on trailer boom lifts) on the base of the machine. The first lever is moving the bottom boom up and down, the second will likewise control the top boom, the third will slew (turn or pivot) the machine on the turntable, and the fourth control is for controlling the jib. (The actual order in which these levers are arranged may change from

FIGURE 3.24 Base controls

machine to machine.) Some machines also have a lever to level the basket of the EWP.

Platform controls

Figure 3.21 shows the platform control module. The levers on this module perform exactly the same functions as those on the base, and will be in the same order.

Deadman pedal

You may be familiar with the **deadman pedal** or **switch** from action movies involving trains or trams. Unless this pedal is fully depressed by means of the operator's foot (in the case of EWP platforms: see Figure 3.22), the machine cannot start or continue to run. The presumption is that it takes a 'live' person to depress the pedal. Should the operator become unconscious, the platform will remain safely static until it is recovered by using the base controls or by the hydraulic bleed valves.

Base/platform override switch

This simple switch is used for directing power to either the base controls or the platform controls. You will have this switched to 'base' or 'chassis' for your pre-operational checks. It should then be switched to 'platform' for the duration of the work period. The switch allows for someone on the ground to lower the machine in an emergency by switching power back to the base control module.

Hydraulic bleed-down valves

These **bleed-down valves** are also known as 'emergency descent valves'. Their purpose is to allow the hydraulic fluid to bleed (drain) out of the pistons and back to the reservoir. Due to the weight of the booms pressing down on the cylinders, this can happen fairly quickly if the valves are opened fully. These valves should not be needed unless the EWP suffers a power failure while the platform is still elevated.

When opening bleed valves, do so slowly and adjust them so the platform descends in a controlled and steady manner. Once power has been restored, make sure the valves are fully closed, then raise and lower the platform at least twice to its full extension using the base controls to ensure that no air has entered the lines or pistons.

Ignition, kill (emergency stop) and fuel switches

There are various forms of, and various places for, these switches. The ignition switch is generally keyed, with separate start and stop buttons (known as a kill switch). The kill switch, also known as the **emergency stop button**, is large so that it can be easily struck in an emergency. (Some will have a twist lock arrangement that requires you to twist and pull to reset it.)

Fuel switches are generally located in the engine compartment on the main fuel line. Except for the fuel, all these switches will be replicated on the platform – that is, you can start and stop the engine from the platform.

Note: There are other means of platform recovery available, and these vary from machine to machine and manufacturer to manufacturer. These include manual jacks to pump rather than bleed hydraulics (in cases where the booms have been articulated over obstacles), manual slewing cranks and backup electric motors.

By now you should have a sound appreciation of what EWPs are, how to determine which one to use, where to locate it, and what the various controls and switches look like and do. We now need to learn how to check a machine to determine if it is safe to use. These checks involve getting into the EWP and operating some of the controls, including raising and lowering the platform. We therefore need to look at the PPE specific to EWPs that may need to be worn.

Individual fall-arrest systems

Individual fall-arrest systems (IFAS – also known as 'fall-arrest harnesses') have been described fully in Chapter 1 of this text. However, as some may choose to read this chapter in isolation, it is pertinent to offer a summarised version here.

Note: All personnel on the platform of an EWP shall wear a correctly fitted IFAS complying with AS/NZS 1891.4 Industrial fall-arrest systems and devices, excluding only scissor and vertical lifts (see 'On-site' box, p. 125). This shall apply irrespective of how high the EWP can, or is intended to, be raised. Personnel on scissor and vertical lifts may be required to wear an IFAS if the risk analysis deems it necessary.

What is an IFAS?

An IFAS consists of the following main components:

- a harness that fits snugly to the body by means of leg and shoulder straps (Figure 3.25)
- a lanyard that connects between the harness and designated mounting point of the EWP platform
- a shock or energy absorber with a maximum deployed length of 1.8 m.

Generally, the energy absorber is integrated into the lanyard (which is then technically called a 'lanyard assembly' – Figure 3.26).

The length of the lanyard is at the discretion of the operator based upon the risk analysis. It should always be as short as possible while still allowing the work to be conducted without undue restraint (which might induce an operator to unclip it from the platform connection point).

The EWPA (2013) recommends the following lanyard lengths:

- for EWP platforms of 1.8 m or greater in width, or when the connection point is at the base of the platform (irrespective of width), use a 1.8 m lanyard
- for EWP platforms less than 1.8 m in width, use a 1.2 m lanyard.

Lanyards may be of rope or web construction. In determining the length of the lanyard assembly, consideration should be given to the overall 'activation clearance' required – in short, how far you will fall before

FIGURE 3.25 Harness

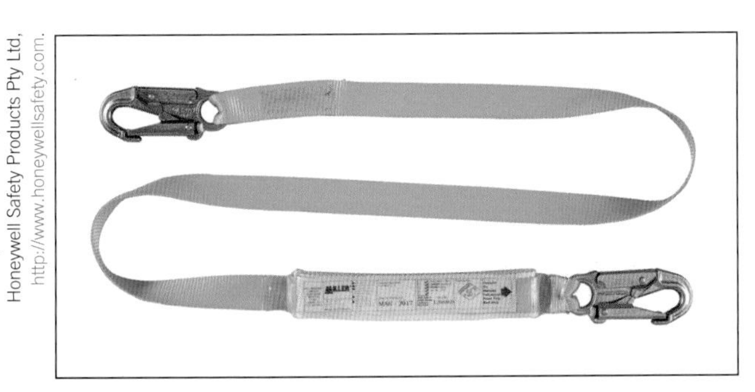

FIGURE 3.26 The energy absorber is integrated into this lanyard

it pulls you to a stop. Figure 3.27 gives the typical activation clearance distance for a lanyard of 2 m in length. Note that you need to have a clear vertical fall distance of 6.5 m from the point of connection: otherwise you are going to hit the ground – hard.

In practice, the fall distance is much reduced in an EWP. This is because the connection point is low within

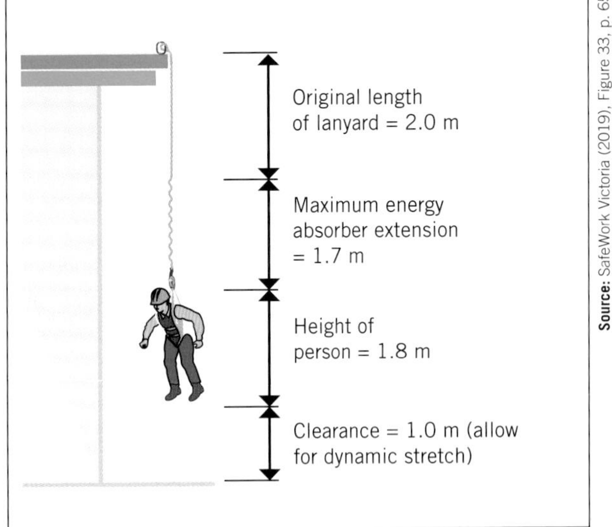

FIGURE 3.27 Activation clearance

the cage. Generally, the activation clearance for EWPs is therefore only 3.5 m or less, depending upon the length of lanyard chosen.

Note: When a fall clearance of less than 3.5 m is not achievable use the shortest lanyard possible and secure it to a platform floor connection points if possible. Alternatively, use a scissor lift or scaffolding if practicable.

All harnesses and lanyard labels must show they have been manufactured to Australian Standards and include an expiry date by which the product should be retired.

Checking your IFAS

Before putting on an IFAS, it is the operator's responsibility to always check that it is in good order first. You are looking for the following:

- the harness, lanyard and shock absorber have been manufactured according to AS/NZS 1891 Industrial fall-arrest systems and devices (Figure 3.28)
- the harness and lanyard assembly have not exceeded their expiry date
- no damage has occurred to the webbing or rope. This may include burns, cuts, rotting, fraying, chemical degradation or crushing
- stitching is sound and unbroken
- buckles, 'D' rings, hooks and the like are not deformed or otherwise damaged
- the energy absorber has not been strained or begun to deploy, and its cover or sheath is intact.

ON-SITE

WHY NOT WEAR A HARNESS IN A SCISSOR OR VERTICAL LIFT?

Wearing a harness and lanyard system in a scissor lift may increase risk of injury through tripping, entanglement and a heightened, but false, sense of security. This can lead to over-reaching or falling against the railing, potentially toppling the lift. It is also argued that a fall from a lift, to which the operator is attached, could topple the machine onto the operator. Likewise, an operator is no longer capable of leaping clear of a

>>

>>

toppling machine but is trapped and again potentially dragged under the machine. Each of these scenarios increases injury potential.

WHEN WOULD YOU WEAR A HARNESS?

Only in instances where access to a point of work requires over-reaching (leaning over the railings, meaning the worker's centre of gravity is potentially outside the safety railing of the EWP). Such instances are to be avoided as a general rule of safety practice, but may occur when no other reasonably practicable action is achievable.

SCISSOR AND VERTICAL LIFT CONNECTION POINTS

Many scissor and vertical lifts have connection points for travel or fall restraint systems only. Restraint systems are designed to limit the travel of the connected worker, 'restraining' them to a limited area (e.g. the EWP platform space). They are designed to stop a person falling in the first place. Restraint lanyards do not have a shock absorber.

An IFAS lanyard should not be connected to these points as they are not designed to withstand the additional shock of a falling person.

HOW TO

FITTING AN IFAS

Figures 3.28 to 3.33 outline the correct procedure for fitting an IFAS. Once completed, it should be a firm fit without being constrictive.

1 Check the harness for any damage and the harness label for adherence to AS/NZS 1891 Industrial fall-arrest systems and devices. Make sure the harness and lanyard assembly have not exceeded their expiry date.
2 Hold the harness by the 'D' ring as shown. Shake the harness and ensure that all the straps fall into place. Make sure the leg and chest strap buckles are unbuckled.
3 Having identified the top rear 'D' ring, put the harness on, much as you would a high-visibility vest. Once the shoulder straps are in position, ensure that the rear 'D' ring is central to your upper back.
4 Adjust and fasten the chest strap.
5 Pull the leg straps between your legs and link to the side buckles as shown. Be sure not to let the straps get crossed or mixed up. Fit should be firm but not restrict movement.
6 Run your hands flat over the webbing to ensure no twisting or bunching has occurred.

Should the harness not feel right, or if there is evidence of incorrect fitting, remove completely and start again.

Note: You must always take ultimate responsibility for ensuring your harness is properly fitted. Whenever possible, however, have a competent person check your harness (particularly the back webbing) for twists and any incorrect connections.

FIGURE 3.28 Check harness labels: step 1

FIGURE 3.29 Harness fitting: step 2

>>

>>

FIGURE 3.30 Harness fitting: step 3

FIGURE 3.32 Harness fitting: step 5

FIGURE 3.31 Harness fitting: step 4

FIGURE 3.33 Harness fitting: step 6, front and back

Source (images): Reproduced with permission of Honeywell Safety Products Pty Ltd, http://www.honeywellsafety.com.

Connecting to the EWP

Having checked and correctly fitted your harness and lanyard, you may now enter the platform of an EWP. On the platform, you will find a lanyard anchorage point (Figure 3.34). *It is to this point only that you must connect.* If the platform is designed for more than one person, then you will find a number of connection points corresponding to the number of persons allowed.

The connection to the EWP is by means of a karabiner or specifically designed hook (Figure 3.35). Only karabiners or hooks requiring a 'double action' to open them may be used. For more information, see 'Hooks and karabiners' in Chapter 1.

FIGURE 3.34 Lanyard anchorage point on EWP platform (circled)

Source: Reproduced with permission of Honeywell Safety Products Pty Ltd, http://www.honeywellsafety.com.

FIGURE 3.35 Karabiner and hook

ON-SITE

A COUPLE OF IMPORTANT PRACTICAL POINTERS

1. Never connect to handrails, midrails or posts. They are not designed for the forces involved in a fall. In addition, the lanyard hook can now slide, and this additional movement may cause the hook to part or break when impacted by the force of a fall.
2. As a general rule, you may only enter or exit an EWP at the support surface level. You may only use an EWP as a means of delivering workers and equipment to an otherwise inaccessible work area when AS 2550.10 is strictly adhered to.

With this knowledge, you may now carry out the necessary pre-operational checks. These will begin without you needing to enter the platform and so may start with the harness off.

KEY POINTS

There are three basic sets of controls to be found on most EWPs:

1. outrigger control module – used to set the machine up level
2. base, ground or chassis control module – used for pre-checks, stowing and emergency control if required
3. platform control module – used during the work procedures.

Other important controls include:

- a keyed switch for selecting either base or platform controls
- a deadman pedal on the platform
- emergency descent systems (bleed-down valves, pumps or backup motors).

In any EWP other than vertical or scissor lifts, you must wear an IFAS harness correctly connected to the platform.

HOW TO

SAFE USE OF MOBILE EWPs

Recently a worker was seriously injured while in a self-propelled scissor lift. The operator was newly trained and had no experience on-site. The task was the application of render on the facade of a new building. However, due to delays, a metal frame feature had already been installed prior to the render being applied. This meant the scissor lift had to be driven between the vertical posts, and under horizontal beams that were just above the height of the EWPs handrail when in the lowered position.

There was no 'spotter' involved – someone to look out for dangerous obstructions such as beams, power lines and the like.

The operator ducked and drove under the beam without checking on the worker applying the render. The worker didn't duck and was crushed between the beam and the EWPs handrail (Figure 3.36).

When working with even seemingly low risk machinery you must remember to check for all potential risks. Use spotters, and check all the occupants/workers that are on board the EWP you are operating. Likewise, if you are not the operator, make sure your communication is clear between yourself and them.

>>

>>

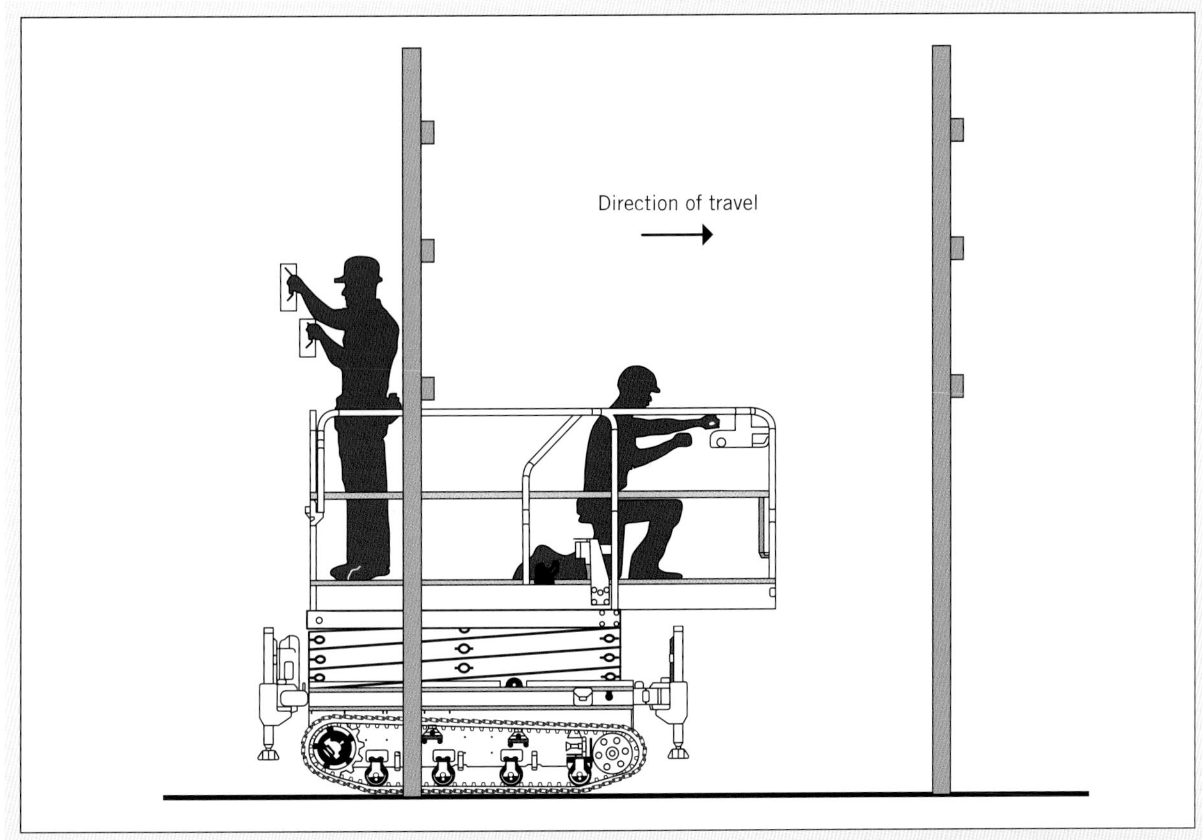

FIGURE 3.36 The operator (kneeling at right) ducked and drove under the beam, the renderer didn't hear the instruction to duck and is about to be crushed between the cross beam and the handrail.

LEARNING TASK 3.4

1. **Circle 'True' or 'False'.**
   An individual fall-arrest system (IFAS) is only required to be worn when you are taking the EWP platform to heights greater than 6 m.
   True False
2. **When choosing an EWP you should:**
   a think *context* as well as task
   b always find the smallest machine available
   c choose a machine that is working *within its operational limits*
   d both 'a' and 'c'

COMPLETE WORKSHEET 3

The safe use of EWPs: Part 2 – checks, loads and positioning

The only way you will really know how to use an EWP is to do just that, use one. Preferably more than one, and more than one type. Experience – experience in multiple contexts, multiple machines and types of machines – is the only road to being a skilful operator.

That said, there are things you still need to learn to put into practice prior to, during and after your first time in an EWP. These include: pre-operational checks; safe loading and position procedures; post-operational checks; and stowing procedures.

Pre-operational checks

At the beginning of each work period, you must run a series of pre-operational checks over an EWP. There are two forms of checks to conduct:

- visual check
- functional check.

The wise operator will do the critical aspects of a visual check every time they get in and out of an EWP. The operational check will be conducted when you first locate the machine.

Pre-operational visual check

The visual check begins with the logbook and then covers the various aspects of the equipment outlined in the following. It is an essential survival skill to develop.

- Logbook:
  - check that there is a current logbook
  - check that the serial number in the logbook matches that on the compliance plate on the machine itself
  - check that all services have been completed and are up to date
  - check that any reported faults have been repaired and signed off.

    If any of the above are negative, the EWP must not be used.
- Operating manual:
  - Locate the manual and check that it is for the correct machine.

    Note: Some manuals cover several similar models. Make sure you are reading the correct sections when looking for information or particular controls on the machine.
- Decals/warning labels (Figure 3.37):
  - Check that any information contained on these match the operating manual and the compliance plate.
  - If you cannot read a label, refer to the manual.
- Fluid levels: The checks needed here will vary. Check the manual for that specific machine for how, where and what to check. Examples include:
  - fuel
  - hydraulics
  - coolant
  - engine oil
  - transmission oil
  - power steering
  - brakes
  - battery.

    Warning: Some fluids must be checked very carefully, as they pose WHS/OHS risks:
  - Hydraulic fluid should not come in contact with your skin (see 'Safety around hydraulics').
  - Coolant can be hot and under pressure within the radiator. Check by looking at the reservoir bottle.
  - Battery fluid is an acid (corrosive). Use safety glasses and chemical-resistant gloves when checking or topping up.
- Major components: Again, the checks vary from machine to machine. In the main, you are looking for cracks and stresses that may weaken a component, rendering the machine unsafe to use. Pay particular attention to welds and points of connection, pivoting or sliding. Examples include:
  - tyres/wheels (including wheel nuts)
  - chassis
  - booms
  - jibs
  - fly booms
  - hydraulic lines and pistons/cylinders
  - oil or other fluid leaks

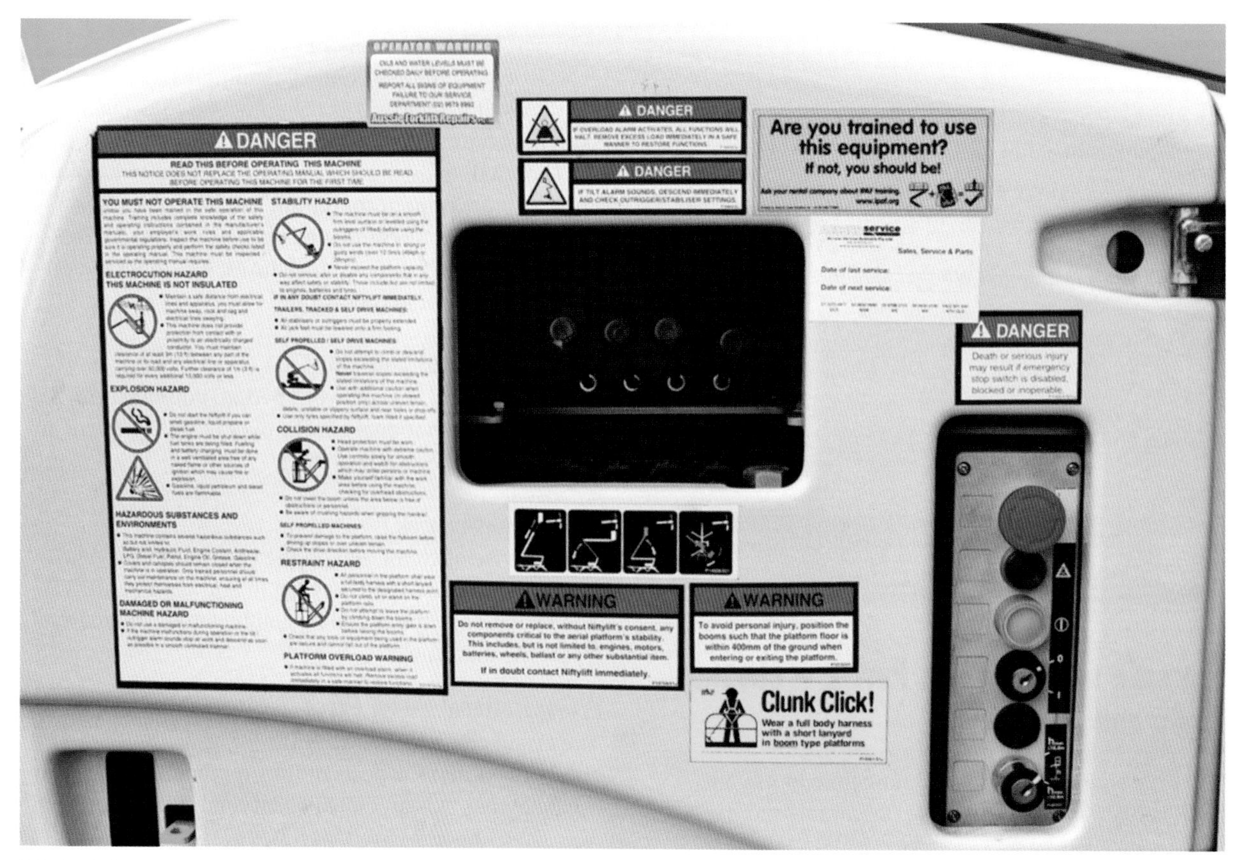

FIGURE 3.37 Decals and warning stickers

- flaking paint around welds
  - turntable and slewing mechanisms
  - scissor arms
  - vertical lift sleeves, chains or booms
  - emergency bleed valves (must be closed)
  - general engine condition
  - outriggers/stabilisers (and all associated components)
  - platform, including handrails
  - controls and switches
  - electrical lines and cables
  - platform gates and IFAS connection points in order.
- Manual brakes are engaged.
- Emergency descent systems: detachable lever arms are where they should be.
- Drive line is engaged (some self-propelled units may be free-wheeled by disengaging the drive line).
- Mains power (240 volts) has been disconnected from charge point (if an electric-driven unit or has a battery backup system).

Pre-operational function check

Depending upon the machine, the pre-operational function check may cover various aspects of the EWP as outlined below. It will end with a logbook entry once you have ascertained that all is well, or if you have found a fault.

Note: Before proceeding with functional checks, *be sure to remove all lock or travel pins*. These hold the booms in position and prevent movement during transit. Failure to remove these pins prior to attempting to raise the boom can lead to damage. In addition, check for slew lock pins (found on some models).

Engine or power off

Check that:

- controls and switches return freely to centre (neutral) when released
- emergency stop buttons function and are released
- keyed switches turn cleanly.

Engine or power on

Where the EWP has outriggers, you will now deploy them as per the manufacturer's manual. While doing so, you must check for any hydraulic fluid leaks that may appear now that the lines have been pressurised. Ensure that the machine is completely level. (Newer machines may have an automatic levelling system – check that it has indeed achieved this. Adjust if necessary.)

Note: In deploying the outriggers, you may need to use road plates or pigsty packing of timbers – see 'Outriggers, road plates and pigsty packing'.

Checking other major functions

You are now ready to check all the other major functions. You will start by checking the base controls.

Base control module

You are looking for clean actions, with no grating or evidence of faulty components. These checks may include:

- the engine start and kill switches work
- the lower boom raises and lowers
- the upper boom raises and lowers
- the fly boom extends and retracts
- the jib raises and lowers
- the slew mechanism turns cleanly left and right
- the scissor arms raise and lower
- the vertical lift components move cleanly
- the platform remains level during all motions
- all audible and visual warning devices are functioning.

Platform control module

At this point of the checks you must fit your IFAS, as you will need to enter the platform. Upon entering the platform, immediately attach the lanyard to the appropriate connection point. *No one may be on the platform without a correctly fitted and connected IFAS*, even for pre-operational checks.

Now repeat all of the above checks using the platform controls. In addition, make sure of the following:

- the engine start and kill switches on the platform work (i.e. they start and stop the engine from the platform)
- the deadman switch is operational: check by lifting your foot while raising or lowering a boom. The boom should stop moving instantly when the deadman switch is released
- the platform remains level during all manoeuvres while loaded.

Emergency descent functions

The hydraulic bleed-down or emergency descent valves have been described previously. These valves should not require testing unless they do not pass your visual inspection (i.e. they look rusted or otherwise unlikely to turn). If you do feel the need to test them for function, you will need to raise and lower the platform twice to its full extension to ensure no air has entered the system. This must be done using the base controls.

Depending upon the design of the particular EWP in service, you may need to check other platform recovery equipment such as:

- manual jacks for pumping hydraulics
- manual slewing cranks
- backup electric motors.

When you have satisfied yourself that the machine is functioning correctly in all aspects, lower the platform, stop the motor, and fill out and sign the appropriate section of the logbook – that is, Section 1: operator safety check record. If you have found a fault, then you must fill out and sign the 'faults' section (Section 2: faults/problems/actions taken). In doing so, you must report it to the owner/hiring agency and/or your supervisor. *Note that the EWP cannot be used.*

Assuming that the machine is functional, you may now load the platform and proceed to work.

Loading and positioning the EWP

Now that the machine is cleared for work, the precautions do not stop. In loading your machine, pay careful attention to the weight of equipment, materials and personnel that you are putting on to the platform. *Never* exceed the stipulated capacity or SWL that is marked on the platform and in the manual. Always load a platform evenly.

In loading the platform, keep in mind the task that you are doing. If it involves removing materials or equipment from heights, then this load needs to be accounted for. For example, if you load to the limit of the platform going up to get something that weighs 20 kg, then you will be 20 kg overweight when you try to come down. If you are working close to the maximum reach or extension of the EWP, this could lead to tipping of the EWP and serious injury or death.

Pay particular attention when loading, or preparing to load, scissor lifts with extending platforms. If you are intending to deploy the extension, limit your total load to the reduced load capacity (where a reduced capacity is stipulated for using the extended deck).

HOW TO

EXAMPLE OF SWL (SAFE WORKING LOAD) CALCULATIONS

- EWP rated at a maximum SWL of 220 kg.
- Two workers/operators each weighing 85 kg, including harness.
- Tools required for task weigh in at 25 kg.
- Materials allowed to be lifted at any one time is found by:

$$\text{Materials allowed} = 220 - [(2 \times 85) + 25] = 220 - 195 = 25 \text{ kg}$$

If you have someone on the platform who is new to working in EWPs, be sure to check their harness and its connection to the platform. Give them clear instructions about limiting aggressive movements while on the platform (such as bouncing, jumping or wrenching).

In positioning the platform ready for work, take time to get a 'feel' for the controls. Try to move the platform with minimal jerking brought about by rapid start or stop movements of the controls. If you have chosen the appropriate machine and located it correctly, you should not need to position the platform at the extremities of the EWP's height or reach. This will make the platform feel more stable to work from.

Also while positioning the platform, take care to monitor your spotter (if required). Whatever means of communication you have chosen to adopt between you, make sure you are listening out for it. Keep your own eye out as well for obstructions and safe working distances. Your spotter is your backup, not the primary means of avoiding dangerous areas: the primary means is *you*.

Once the platform is raised into position, you should turn off the engine or electric motor to conserve fuel (or power). The platform will remain static while you do your work. Note that if it does not remain static, and you notice a slow lowering of the platform over time, then there is a fault in the hydraulics and the platform must be immediately lowered. You must then:

- record the fault in the log
- inform the owner and/or supervisor
- bar the machine from use until the fault is found and repaired.

Emergency descent procedures

There are a number of occasions when you may have to use the emergency descent procedures to lower a platform and its occupants safely to the support level. Not all instances seemingly qualify as an 'emergency', but all should be treated as such. The following are examples of emergency situations and the procedures that should be undertaken in each case.

Motor failure

If the motor fails while you are operating the machine from the platform and will not restart, the procedure is as follows:

1. If a spotter is in attendance, request that they attempt to restart the motor from the base control, check fuel levels and the like.
2. Assuming the motor doesn't restart, lower the machine using the emergency descent valves on the platform.
3. The spotter may be required to use manual slew and/or pump devices to clear obstacles.

Operator injury

If the spotter notices that the operator is injured, or has fainted or otherwise collapsed, and is not responding to repeated calls, the procedure is as follows:

1. Do not touch the EWP until you are certain that the machine has not entered the exclusion zone of a power supply line.

   If the EWP is clear of the exclusion zone of power supply lines:
2. Immediately lower the platform by means of the base control module if the motor is running or can be started.
3. If the motor won't start, immediately lower the platform by means of the emergency descent valves, using pumps and/or manual slew devices as required.
4. Carry out emergency first-aid procedures as required by that work site or your company's policy (i.e.

contact the emergency services and carry out appropriate first aid or get a first-aid officer to the injured person as quickly as possible).

If the EWP is within a power supply exclusion zone, or has made contact with power lines:

1 Do not touch or approach the EWP.
2 Immediately contact the emergency services and/or, where required, the site supervisor and site emergency services.
3 Stay on-site to be of assistance to the emergency services as required.

The EWP develops a lean or the platform becomes unlevel during operation

In this case, the procedure is as follows:

1 Immediately lower the platform by means of the platform control module, assuming the motor is running or can be started.
2 The owner or site supervisor must be notified and the EWP must be removed from service until repaired.

The EWP platform does not stop moving when you let go of the control levers or return them to the stop position

In this case, the procedure is as follows:

1 Immediately lower the platform by means of the platform control module, assuming the motor is running or can be started, and that the controls will function adequately for you to do so.
2 If controls cannot be used adequately, lower the platform by means of the emergency descent valves.
3 The owner or site supervisor must be notified and the EWP removed from service until repaired.

Lowering, stowing and/or parking an EWP

Lowering an EWP is no different from positioning it in the first instance. Switch on the engine and start your descent following the same safety procedures you followed on the way up. Keep your mind on obstacles, crush hazards and the like. You may require greater feedback from your spotter as you will be unable to see obstacles beneath you from the platform.

If you have gone up to retrieve something, the platform will now be heavier than it was on the way up. Be more particular in the use of the controls.

If you are using a self-propelled EWP, you will need to park it somewhere. Consider the following when doing so:

- paths of access and egress from the site/work area
- where you need to refuel or energise
- vehicular and pedestrian traffic
- fire and emergency exits
- location of first-aid facilities
- hazardous materials
- railway and/or tram tracks
- power lines.

If you are using a boom lift, be sure to set the boom down into the rest cradle and lower the jib or platform completely to allow for ease of exiting the machine. *You may only exit the platform of an EWP once it is at its lowest rest position.* Once you are down, and the equipment, tools and/or materials have been removed from the platform, you are ready to stow the machine for transport.

For machines with stabilisers or outriggers, the procedure for stowing is the reverse of that for deployment:

- raise the outriggers back to their stow position (if fitted) – insert travel pins if fitted
- switch off the engine
- remove the key
- insert all boom and/or stabiliser lock pins.

If the EWP is to be used the following work day, it is not advised to stow the machine in an elevated condition. Only do so if your risk assessment suggests sound reasons for doing so. High winds are not always predictable and could create a dangerous condition, stress the booms, or even topple the machine.

You are now ready to carry out your post-operational checks prior to completing the logbook entry.

Post-operational checks

Given that you have just been using the machine (and assuming you have found no faults in the process, nor damaged it in the descent), the post-operational check is visual only. Much like the pre-operational visual check, look to the following:

- tyres/wheels (including wheel nuts)
- chassis
- booms
- jibs
- fly booms
- hydraulic lines and pistons/cylinders
- oil or other fluid leaks
- turntable and slewing mechanisms
- scissor arms
- vertical lift sleeves, chains or booms
- controls and switches
- electrical lines and cables
- travel pins (boom, slew, outrigger lock pins) in place
- keys removed.

In a post-operational check you are paying particular attention to booms, and points of possible stress (pivot points, hydraulic arm connection points and the like). You are looking for fractures, stresses to weld, cracks or delaminating of fibreglass on insulated booms, and so on.

If any faults are found, you must follow the fault-finding procedure – that is, record the fault(s) in the logbook, and contact the supervisor, who will declare the machine unfit for use until repaired.

Assuming all is OK, fill in the logbook: state the time concluded and the hours worked, note that the post-operational check has been completed successfully, then sign and date it.

Leaving the site

Having completed the tasks and stowed the EWP, or organised its removal from the site, the 'no go' area is ready to be reopened. On large sites you must contact the site supervisor prior to leaving the site or removing barriers, as only this person (or their delegate) may open the site to other personnel or traffic.

KEY POINTS

- Pre-operational visual checks ensure that all elements of the equipment appear to be ready for use (welds, cracks, hydraulic leaks, etc.), and for use for the duration of the task (fuel, oils, batteries, etc.).
- Pre-operational function checks ensure all elements of the machine work as they should; for example, hoses don't leak when under pressure, no cracks become visible when under load.
- When calculating SWLs, you must include all loads applied to the machine, including people, tools and all materials.
- When positioning the platform, use the controls smoothly, pay attention to your spotter and/or possible obstructions. Locate the platform so that you do not have to stretch or reach excessively to do the task.
- Generally, stow or park an EWP in the down position rather than up and ensure all travel pins are in place (where required).
- Fill out the logbook after post-operational checks.

Outriggers, road plates and pigsty packing

Outriggers or stabilisers are designed to increase the effective width of a machine and so increase its stability when a platform and load is raised. They are particularly important on uneven, broken, sloping or unstable ground. Deploying outriggers appropriately is not always easy in any of these cases, however. To ensure that each outrigger is taking up an equal share of the load, **road plates**, and/or timbers applied in a manner known as **pigsty packing**, are often required.

Road plates are metal or thick-ply sheets designed to distribute the load of the much smaller outrigger footplate over a larger ground area. The size of a road plate may vary with the need, though some larger truck-mounted EWPs carry plates as standard equipment. Road plates are effective only when laid reasonably level, so the ground needs first to be levelled off by digging or scraping, or by the addition of timbers placed underneath.

Pigsty packing is a collection of large section timbers (generally 100–200 mm × 45 mm) which are stacked on each other, as shown in Figure 3.38. The advantage of pigsty packing over road plates is that they are lighter, and less earth needs to be removed to create a stable, level surface. The 'downside', however, is that rather than just four road plates, you need quite a number of pieces of timber, which can be bulky to store. Nevertheless, most truck-mounted EWPs will carry a number of timbers for this purpose, even if they are carrying road plates as well.

FIGURE 3.38 Pigsty packing

When laying pigsty packing, be sure to begin by levelling the earth (with a shovel or by simply scratching at the ground with the end of the timber) under each of the first timbers. On uneven or sloping ground, it is permissible to have the first timbers at different levels, as long as those that follow will be level. Pigsty packing can be used in conjunction with road plates to reduce the risk of the outrigger footplates separating the timbers.

LEARNING TASK 3.5

1 **Circle 'True' or 'False'.**
When calculating safe working loads (SWL) you must *include all loads applied to the machine*, including people, all tools and all materials.
True False

2 **Pigsty packing:**
- a cannot be used on sloping ground
- b must not be used in conjunction with road plates
- c can be applied to sloping ground by having the first timbers at different levels, making the next layer level
- d can only be used when a road plate is levelled into the ground first

GREEN TIP

An EWP is designed to get you close to the work location in often dangerous, high spaces. Always consider the environment around the machine and seek to do no harm to it by you being there for your work. Don't leave machines running for longer times than necessary. Don't remove more foliage than is necessary. Repair ground surfaces you may have damaged.

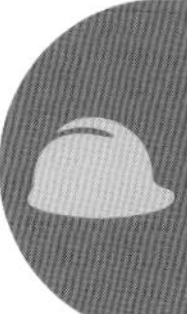

It is vital that you obtain the EWP instruction manual and training from a competent person before you attempt to operate a machine. Never try to figure out how to operate an EWP without guidance, as these machines can be dangerous.

COMPLETE WORKSHEET 4

Clean the work area

Once the work has been completed, it's important to clean the work area.

Any material that was left over from the job should be reused where possible and put away. Any scrap materials should be recycled or put into waste disposal in accordance with workplace policies and procedures.

As noted previously, post-operational checks on the EWP need to be undertaken while ensuring the plant is left clean and is ready for the next person to use.

Tools, PPE and other equipment used during the task should be checked for any signs of damage and maintained and stored in accordance with the manufacturer's instructions.

The work supervisor should be informed that the works have been completed and the area has been restored to its pre-work condition.

SUMMARY

As stated at the beginning of this chapter, the safe operation of an EWP requires not only extensive knowledge of the equipment and procedures, but also experience in the context. Whenever possible, work with an experienced operator who can act as a mentor and guide. The wisdom of experience can never be understated with regard to equipment of this nature. This is particularly so when working around power lines and/or trees, or when the EWP is set up on sloping ground.

The aim of this chapter has been to help you develop the knowledge base with which to informatively engage with each context and with those who are experienced and helping you along the way. The key areas covered include:

- There are various types and designs of EWPs and where you might use them. Remember to always select a machine that is working well within its limits, rather than at full reach. When required to work near live power lines, be sure to use an EWP with an insulated boom.
- Logbooks and manuals must be checked and filled out prior to and after use. You are looking for proof that the machine is fit for service and that all maintenance records are complete.
- Hydraulics is an often-forgotten element of EWP training and use. Do not underestimate the importance of this element of your machine and the associated risks. Carry out your checks on the hose lines with care and as instructed.
- The safe use of EWPs was dealt with on two levels. The first covered the importance of spotters, hazard identification and the actions of each of the controls – the 'job planning'. The second part dealt with pre- and post-operational checks, locating the correct machine for the context and, finally, how to safely stow it for transit.

It is important that you learn to carry out all of these actions and checks with rigour. Test the machine by using the ground controls and take nothing for granted, particularly when checking hydraulic lines and seals. Ultimately, your safety and that of others, both on the ground and in the basket, is totally dependent upon you and your diligence. Act accordingly.

REFERENCES AND FURTHER READING

Text

Australian and New Zealand Resuscitation Council (ANZCOR) (2021), ANZCOR Guideline 9.1.5 - First Aid Management of Harness Suspension Trauma, ANZCOR, East Melbourne, VIC, **https://resus.org.au/download/anzcor-guideline-9-1-5-harness-suspension-trauma-first-aid-management-april-2021-0-3-mib/?wpdmdl=13734&masterkey**

Elevating Work Platform Association of Australia (EWPA) (2018), Information Sheet: Policy on the use of Fall Arrest Systems in Elevating Work Platforms v2.0 April 2018, EWPA, Mona Vale, NSW, **https://ewpa.com.au/wp-content/uploads/2020/04/24.-Policy-on-the-Use-of-Fall-Arrest-Systems.pdf**

Elevating Work Platform Association of Australia (EWPA) (2019), Information Sheet: *Distance from Power Lines*, EWPA, Mona Vale, NSW, **https://ewpa.com.au/wp-content/uploads/2020/10/Distance-from-Power-Lines-July-2013.pdf**

Elevating Work Platform Association of Australia (EWPA) (2022), Good Practice Guide: Mobile Elevating Work Platforms v1.03 June 2022, EWPA, Mona Vale, NSW, **https://ewpa.com.au/resources/good-practice-guide/**

Safe Work Australia (2020), *Managing the Risk of Falls at Workplaces: Code of Practice*, Commonwealth of Australia, Canberra, **https://www.safeworkaustralia.gov.au/sites/default/files/2022-10/Model%20Code%20of%20Practice%20-%20Managing%20the%20Risk%20of%20Falls%20at%20Workplaces%2021102022_0.pdf**

WorkSafe Victoria (2019), *Prevention of Falls in General Construction: Compliance Code*, 2nd edn, WorkSafe Victoria, Melbourne, **https://content.api.worksafe.vic.gov.au/sites/default/files/2019-12/Compliance_Code%3A_Prevention_of_falls_in_general_construction.pdf**

Online resources

Elevating Work Platform Association of Australia (EWPA), **http://www.ewpa.com.au**

Workplace Health and Safety Queensland Electrical Safety Office, **http://www.worksafe.qld.gov.au**

Relevant Australian Standards
AS 2550.1 Cranes, hoists and winches – Safe use – General requirements
AS/NZS Cranes, hoists and winches Part 10: Mobile elevating work platforms
AS 2550 Cranes, hoists and winches – Safe use set
AS 2550.10 Cranes, hoists and winches – Safe use – Mobile elevating work platforms
AS 2550.10 (Amdt 1) Cranes, hoists and winches – Safe use – Mobile elevating work platforms
AS/NZS 1891 Industrial fall-arrest systems and devices

GET IT RIGHT

Source: JL Hasted/Office of Industrial Relations, Workplace Health and Safety Queensland

FIGURE 3.39 Poor 'pigstying' of timber supports

In Figure 3.39, the outrigger on an EWP has been packed, as it cannot be lowered sufficiently to engage with the ground. Based upon what you have learnt in this chapter, identify the preferred manner in which this should be done.

___

___

___

WORKSHEET 1

Student name: ______________________

Enrolment year: ______________________

Class code: ______________________

Competency name/Number: ______________________

To be completed by teachers	
Student competent	☐
Student not yet competent	☐

CPCCCM3001: Operate elevated work platforms up to 11 metres – specifically, the element 'Plan and prepare'.

Task: Answer the following questions.

1 What is an EWP (also known as a MEWP)?

____

____

____

2 List the five categories of EWP.

1 ______________________

2 ______________________

3 ______________________

4 ______________________

5 ______________________

3 Circle 'True' or 'False'.

Having completed the competency CPCCCM3001 you are allowed to operate any height capacity boom-type machine provided you do not take it up beyond 11 vertical metres above the floor or ground surface. True False

4 Circle 'True' or 'False'.

To work in a boom-type EWP at heights greater than 11 metres you are required to hold a High Risk Work Licence. This licence allows you to operate all types of EWPs including scissor and vertical lifts. True False

5 Which Australian Standard applies to EWPs?

____

____

6 Modifying an EWP includes:

a Drilling a hole in the engine cover
b Putting in a screw to hang keys off
c Cutting into or welding an extra section onto the main boom
d All of the above

7 After an EWP has been modified, the following must be supplied to the operator:

a Evidence of approval and compliance
b Photographs of the modifications made
c The name of the person who actually did the welding
d All of the above

8 Circle 'True' or 'False'.

A fly boom is incorporated into boom lifts to extend their reach.	True	False

9 If you decide that you need to modify an EWP, you:

a Can do so if the modifications are minor and you enter them in the logbook before use
b Must contact and gain approval through your relevant state or territory WHS/OHS authority
c Must contact and gain approval through the supplier of the machine
d Have chosen the wrong machine as you cannot make any modifications to a registered EWP

10 Your vehicle weighs 2500 kg. What is the maximum weight of a braked trailer mounted EWP that you can tow?

a 4000 kg
b 1667 kg
c 3750 kg
d 2500 kg

11 When intending to modify the control unit and/or add baskets to the platform on an EWP you must:

a Notify the manufacturer and supply them with details after completion
b Notify your relevant state WHS/OHS authority before doing any modifications
c Ensure the new load and reach limits remain within the manufacturer's specifications
d Stop! These modifications are prohibited

12 Circle 'True' or 'False'.

Compliance plates are a critical element of an EWP; having the logbook with model number, date of manufacture and SWL is not sufficient to begin the safety checks or operate the machine.	True	False

13 Truck-mounted EWPs have insulated booms so that the:

a Hydraulic hoses do not freeze or overheat in extreme outdoor conditions
b Truck motor vibrations do not carry to the platform
c Boom cannot carry electricity to ground
d Noise of passing vehicles or nearby machines cannot vibrate the platform when fully extended

14 List the two main modifications that must never be attempted:

1 ______________________________

2 ______________________________

15 Circle 'True' or 'False'.

Compliance plates are not a critical element of an EWP, as long as you have the logbook with model number, date of manufacture and SWL. True False

WORKSHEET 2

To be completed by teachers	
Student competent	☐
Student not yet competent	☐

Student name: ______________________

Enrolment year: ______________________

Class code: ______________________

Competency name/Number: ______________________

CPCCCM3001: Operate elevated work platforms up to 11 metres – specifically, the element 'Conduct routine checks of EWP'.

Task: Answer the following questions.

1 An EWP logbook has three main parts and a preface. What information will you find in the preface?

____

2 What are the three main parts of a logbook?

1 ______________________

2 ______________________

3 ______________________

3 Circle 'True' or 'False'.

The logbook for an EWP should always be kept in a weatherproof pouch attached to the machine. True False

4 The logbook of an EWP provides the operator or inspectors with information and a means of documenting events. List three of these things that the logbook is used for:

1 ______________________

2 ______________________

3 ______________________

5 Before using the EWP you must check Parts 2 and 3 of the logbook for what information?

____

____

____

6 The pre- and post-operational checklists are generally found in what part of the logbook?

____

____

7 When logging a fault on an EWP, what information must you enter into the log and in what part?

____

____

____

8 List four items of important information that you will normally find in the operating manual of an EWP.

1 ______________________________

2 ______________________________

3 ______________________________

4 ______________________________

9 If you are unfamiliar with a machine, what should you do?

a Read the operating manual and be tested on it by the local authority.

b Not use the machine, but instead find another machine with which you are familiar.

c Contact the owner or your supervisor to organise familiarisation training and get a copy of the manual.

d Use the machine, but do not take it to its full height for reach.

10 List five key characteristics of hydraulic oil.

1 ______________________________

2 ______________________________

3 ______________________________

4 ______________________________

5 ______________________________

11 You notice a small bulge in a hydraulic hose, but it is not leaking. You should:

a Contact your supervisor, the service agent or hiring company and get it replaced before commencing work

b Fill out the faults section of the logbook, complete the work and ensure the service agent or hire service are made aware of the issue

c Continue to do the work and let your supervisor or hire agent know afterwards

d Take no action as it still works and the hire agent or supervisor will deal with it if it becomes serious later

12 List four precautions needed when checking hoses, lines and pistons.

1 ______________________________

2 ______________________________

3 ______________________________

4 ______________________________

13 Can you attempt to repair, tighten or adjust hydraulic hoses or couplings yourself? Explain your answer.

____

____

____

14 Before using an EWP you must fill out which section of the logbook?

____

WORKSHEET 3

To be completed by teachers	
Student competent	☐
Student not yet competent	☐

Student name: ______________________________

Enrolment year: ______________________________

Class code: ______________________________

Competency name/Number: ______________________________

CPCCCM3001: Operate elevated work platforms up to 11 metres – specifically, the elements 'Locate EWP in place' and 'Elevate platform to work location'.

Task: Answer the following questions.

1 What is the purpose of a 'spotter' with respect to EWP operation?

____

____

2 Why might you need more feedback from your spotter when lowering the platform of an EWP than when raising it?

____

____

____

3 List two methods of communication that a spotter may use as a backup to radios.

1 ______________________________

2 ______________________________

4 Describe what is meant by 'context' when it comes to choosing an EWP.

____

____

5 Choosing an EWP that is working well within its limits means:

a You are working within your own limits, both physically and mentally

b It puts less wear on the machine

c It reduces the chances of accidents through overreaching or overloading

d All of the above

6 List four vital responsibilities of a spotter.

1 ______

2 ______

3 ______

4 ______

7 List three possible considerations you may have to make with regard to people and places outside of the immediate work area.

1 ______

2 ______

3 ______

8 The lanyard used to connect your harness to the EWP must have a hook or karabiner with what type of action?

______

9 Why is it best to work from a platform that is not at the full extremities of its reach?

______

______

______

10 List five authorities to which you might need to apply for a permit to undertake the planned works.

1 ______

2 ______

3 ______

4 ______

5 ______

11 How high off the support surface must an EWP platform be before an IFAS can safely be deployed?

______

12 Where will you find information about the wind rating (maximum wind speed) for which you can safely operate a particular EWP machine?

______

______

______

______

13 The connection points of some scissor and vertical lifts are load limited to what type of lanyard?

____

14 You find the lanyard on your IFAS is limiting your reach to the work you are doing from an EWP. Can you undo it or temporarily attach it around a handrail for the short period of time you need to reach that awkward spot? Explain why or why not.

____

____

15 Explain the difference between an 'exclusion zone' and a 'no go' area.

____

____

____

16 The safe working distance from power lines before you must apply for written permission from your local electrical authority (as recommended by the Elevating Work Platform Association of Australia) is:

a 6.4 m (standard poles) and 10 m (towers)

b 3 m (standard poles) and 6.4 m (towers)

c 6.4 m (standard poles) and 8 m (towers)

d 3 m (standard poles) and 8 m (towers)

17 Five hazards that you are likely to encounter when using an EWP are described in the chapter. What are the risks associated with each of these?

1 ____

2 ____

3 ____

4 ____

5 ____

18 The best location for the EWP will have been identified during the creation of your SWMS or JSA documents. In so doing, you will have picked a spot that has minimal reliance upon:

1 ____

2 ____

WORKSHEET 4

To be completed by teachers	
Student competent	☐
Student not yet competent	☐

Student name: ______________________

Enrolment year: ______________________

Class code: ______________________

Competency name/Number: ______________________

CPCCCM3001: Operate elevated work platforms up to 11 metres – specifically, the elements 'Lower platform and shut down' and 'Clean up'.

Task: Answer the following questions.

1 An EWP has an SWL (safe working load) of 280 kg. You, with your tool belt, weigh 85 kg, and your co-worker, with their belt, weighs 90 kg. The tools you need weigh another 25 kg. What is the total weight of materials that you may take up and/or bring down with you?

a 65 kg

b 70 kg

c 80 kg

d 95 kg

2 One form of pre-operational check is visual. Name the other form.

____

3 You and a co-worker must complete a range of tasks on the ceiling of a light commercial building. You will use an EWP that has an SWL of 250 kg. You and your co-worker have a combined weight of 190 kg. The tools, equipment and materials weigh in at 50 kg. You will affix to the job 15 kg of these materials, but have to bring down an air-conditioner unit weighing 35 kg. The air-conditioner unit only requires some minor tools to remove, but both you and your co-worker are needed to handle it. How might you safely perform this task?

____

____

4 List three fluids that must be checked with great care, and state why.

1 ______________________

2 ______________________

3 ______________________

WORKSHEETS 3

5 When checking major components of an EWP, what are you looking for in particular?

____

____

____

____

____

6 The EWP has outriggers which you deployed. Now they are positioned, and the EWP is sitting level, you must:

a Check that the outriggers are all fully extended
b Check for any hydraulic fluid leaks as the system is now pressurised
c Fill out the logbook to confirm that the outriggers are operable
d Have your spotter confirm that the machine is correctly positioned

7 What must you do before proceeding with the pre-operational functional checks?

____

____

8 Circle 'True' or 'False'.

Initial testing of an EWPs main actions (raising and lowering or extending booms, lifts and the like) is undertaken using the base control module. True False

9 What must you do before you conduct the platform control module tests?

a Fill out the logbook with regards to the outrigger performance.
b Put on your IFAS harness.
c Switch the engine off.
d Put on you IFAS harness and immediately connect your lanyard to the platform connection point upon entry.

10 Aside from repeating the functional checks you made using the base controls, list three further checks you must do using the platform controls.

1 ______________________________

2 ______________________________

3 ______________________________

11 You decide to check the emergency descent valves for function during the pre-operational checks. Having closed the valves, you must raise and then lower the machine to its maximum extent to ensure there are no leaks. How will you conduct this test?

____

____

____

12 List three emergency features (other than the descent valves) that you might have to check, depending upon the EWP that is in service.

1 ___

2 ___

3 ___

13 You are the spotter for a worker using an EWP. You notice that the worker has dropped back into the platform and is not responding to your repeated calls. Which of the following actions do you take?

a Immediately lower the platform using the emergency descent valves.

b Immediately lower the platform using the base control module.

c Check that the EWP has not entered an electrical supply exclusion zone before lowering the platform using the base control module.

d Contact the emergency services and await their arrival before lowering the platform.

14 On occasion, you may have gone up in an EWP to retrieve or remove an item or a piece of equipment. What implication does this have for lowering the platform compared to raising it?

___

___

___

___

15 You may need to 'park' a self-propelled EWP ready for use in the following days on a site. What should you consider when choosing your parking spot?

___

___

___

___

16 As a general rule you may exit an EWP:

a When the platform is positioned against any solid surface such as a scaffold or upper level floor

b When the platform is at the support surface level

c When there are no high winds impacting the machine

d When the spotter indicates that it is safe to do so

17 Under what conditions may you use an EWP to deliver workers and/or tools and equipment to otherwise inaccessible work areas?

___

___

18 When lowering a boom-type EWP, what must you be sure to do?

____

____

____

19 List the four steps in stowing an EWP.

1 ______________________________

2 ______________________________

3 ______________________________

4 ______________________________

20 It is not recommended to stow an EWP in the fully raised position because:

a It adds pressure to the hydraulic system

b The stabilisers are not designed to hold loads for extended periods of time

c High winds may strain the main components and or topple the machine

d (a) and (b)

21 The post-operational check is generally visual only. What are you looking for in particular?

____

____

____

____

____

22 You are repositioning the EWP platform, when the motor stops and will not restart. What do you do?

____

____

23 Having positioned the EWP platform, what should you do?

____

____

24 Having positioned the EWP platform, you notice a slight lowering of the platform over time, or that the platform becomes unlevel. What must you do?

____

____

____

____

25 What is a road plate? What is its purpose?

What it is: ______________________________

____

Purpose: ______________________________

____

26 Describe the process of laying pigsty packing.

____

____

27 When positioning the EWP platform ready for work, you should limit jerky, stop-start use of the controls. Why?

____

____

28 When working from the platform of an EWP, what actions should you avoid?

____

____

____

LIMITED-HEIGHT SCAFFOLDING

4

This chapter covers the following topics from the competency 'Limited-height scaffolding':

1 Background
2 Scaffolding: what is it, and why do we need it?
3 Planning the job: hazard assessment
4 Choosing the right scaffold
5 Working platforms and duty ratings
6 Scaffold assembly: preliminaries
7 Modular scaffold
8 Frame scaffold
9 Bracket scaffold
10 Trestle scaffold
11 Tower frame scaffold
12 Tube-and-coupler components: ties and other uses
13 Inspecting and maintaining the scaffold
14 Rope, knots, gin wheels and handballing
15 Consolidation: creating a SWMS document

Unlike the previous chapters in this text, this chapter doesn't cover a 'topic' as such. This is because scaffolding should be considered more rightly as a trade unto itself. Whole books have been written on the subject, and years of broad-based experience on both commercial and domestic sites is needed before anyone could claim to be truly a 'scaffolder' – this in addition to any training and qualifications that may be obtained along the way.

In light of this, the material within the chapter focuses upon those areas directly of relevance to the construction of limited-height scaffolds and the appropriate unit of competency. An overview of the chapter and expected learning outcomes is provided below.

Overview

This chapter addresses the key elements of the following unit:

- CPCCCM2008: Erect and dismantle restricted height scaffolding

Prerequisites

- CPCCWHS2001: Apply WHS requirements, policies and procedures in the construction industry

***Important note:* Undertaking this unit *does not* lead to a Basic Scaffolding licence, or any other licence. See 'Background' on the next page.**

Background

As a trainee, you should be aware from the outset that upon successful completion, *you can only construct scaffolds to a maximum working platform height of 4 m*. To go beyond this height, further training, experience and qualifications will be required; in other words, the unit you are undertaking does not lead to a Basic Scaffolding licence, or any other licence. It provides you only with the requisite skills and knowledge to assemble a safe scaffold to 4 m (for which no licence is required). To construct a scaffold above this 4 m limit without a recognised qualification is not only a breach of regulations, but may also put your life, and the lives of others, seriously at risk.

So why do this unit?

Scaffolding is very much a part of both the domestic and commercial construction scene. The knowledge and skill of selecting, locating, constructing and maintaining appropriate and safe working scaffolds is essential to almost all of the many and varied construction trades: be it painting and decorating, carpentry, bricklaying, plastering, or any other.

Even if you don't have to build the scaffold yourself, safety is everyone's business, so you need to be able to identify a safe scaffold from one in need of repair, further support or reconstruction.

To this end, the chapter will take you through the basics of bracket, unit frame, modular and lightweight scaffolds, as well as the relevant application of tube-and-coupler components, ladders, trestles, ropes and knots. In so doing, it offers the information required to construct each of these types of scaffolds, as either free-standing or tied-in structures, to a height of 4 m. To begin with, we shall take a broad look at just what scaffolding is and why we use it.

KEY POINTS

- This unit of study does not lead to a Basic Scaffolding licence.
- Upon completion of this unit, you can construct scaffold to a maximum of 4 m without a licence.

Scaffolding: what is it, and why do we need it?

Borrowing from the Australian Standard (AS) AS/NZS 1576.1 Scaffolding – Part 1: General requirements, scaffolding may be defined as:

A temporary structure that generally includes one or more working platforms, access platforms, catch platforms and/or landing platforms.

While from AS 4576 Guidelines for scaffolding, a scaffolder may be regarded as:

Someone engaged in erecting, altering or dismantling scaffolding.

Unfortunately, the definition of scaffolding is somewhat vague and unhelpful if you don't already know what working platforms, **access platforms**, catch platforms or **landing** platforms actually are. Nor does it cover the issues of competence or the need for a licence for particular scaffolding activities.

The key term in the definition is the word 'temporary'. Scaffolding is generally a lightweight, yet strong, structure designed and positioned to allow other work to be done. It is then removed or repositioned. Figures 4.1 to 4.9 give you some idea of how broad the definition of scaffolding must be, as well as offering some clear examples of common scaffolding types and applications.

FIGURE 4.1 Modular scaffold

FIGURE 4.2 Tube-and-coupler scaffold

Source: Images by GCS

FIGURE 4.3 Access platforms

FIGURE 4.4 Bracket scaffolding

Some scaffolding is in fact 'permanent', in as much as a **cradle** may be permanent equipment on a high-rise building – such scaffolds are used for cleaning, maintenance or repair work. However, they also fit the definition of temporary, in that they are moved and repositioned (raised and lowered) as the need arises.

Scaffolding is not limited to those types shown in Figures 4.1 to 4.9, however. Other types include mast climbing platforms, spur scaffolds, bracket scaffolds, and the list goes on, with AS/NZS 1576.1 identifying over 25 distinct types. Given the scope of the competency addressed, this chapter shall focus on only five types of scaffold:

1. modular scaffold
2. frame scaffold
3. bracket scaffold
4. trestle scaffold
5. tower frame scaffold.

In addition, we shall discuss issues of stability controls, such as tying in your scaffold to existing structures, and **raking shores**.

Source: Image courtesy of WernerCo

FIGURE 4.5 Mobile scaffolding

Source: Images by GCS

FIGURE 4.6 Suspended scaffolding

Source: Shutterstock.com/Aisyaqilumar2

FIGURE 4.7 Birdcage scaffold

Source: Safe High-Ts Australia Pty. Ltd.

FIGURE 4.8 Frame scaffold

Source: Dreamstime.com/Eric Flamant

FIGURE 4.10 Safety nets

Source: Image courtesy of WernerCo

FIGURE 4.9 Tower frame scaffold

Source: Images by GCS

FIGURE 4.11 Catch platforms

The role of scaffolding in construction

The need for scaffolding is fairly clear from the accompanying images, let alone from your own experience in, or observations of, the construction industry. Good scaffolding allows us to carry out various work activities safely at heights. It does so by the creation of a solid, stable structure known as a 'working platform'. Scaffolding can also be used for, or as:

- the temporary location of building materials (bricks, mortar, paint, timber and the like)
- a safety barrier around areas where a fall of greater than 2 m is possible
- the positioning of safety nets (Figure 4.10) and catch platforms (Figure 4.11)
- a means of accessing elevated work areas (see Figure 4.3).

As has been explained in Chapter 1 of this text, working at heights is an area of constant focus in the development of work health and safety (WHS; known in Victoria as occupational health and safety, or OHS) regulations. This is because it is the area in which far too many deaths and injuries occur in the industry. Chapter 1 also outlined a concept known as the 'hierarchy of control', which we also discuss later in the chapter:

- Level 1: Undertake the work on the ground or on solid construction.
- Level 2: Undertake the work using a passive fall prevention device.
- Level 3: Undertake the work using a work positioning system.
- Level 4: Undertake the work using a fall-arrest system.

Appropriate scaffolding is a level 2 control. This means that, short of bringing the work to the ground, or building a solid floor or platform to work on, this is considered the best means of providing safe access to elevated work areas. This remains so even in cases where the construction of the scaffold itself would require a level 4 control such as a fall-arrest system.*

The critical question of what is an 'appropriate' scaffold is discussed below under the headings 'Planning the job: hazard assessment' and 'Choosing the right scaffold'. Before we consider those topics, it is worth looking at who can build a scaffold.

Who can build a scaffold?

Chapter 1 outlines the history and application of the national *Model Work Health and Safety Act* (Model WHS Act; Safe Work Australia, 2019a) and its implications for the WHS or OHS (Victoria) Acts of each state and territory. Codes of practice, compliance codes and guidance material guide organisations and their workers on appropriate safe work practices.

In the case of scaffolding over 4 m in height, this path of regulation has led to the *National Standard for Licensing Persons Performing High Risk Work* (Australian Safety and Compensation Council, 2006) and (for those states and territories that have adopted the Model WHS Act) the Model WHS Regulations (Safe Work Australia, 2019b). Each effectively describes the same three-level licensing system:

1. *Basic Scaffolding:* Covers **prefabricated scaffolds** (e.g. modular and frame scaffolds), bracket scaffolds, gin wheels, safety nets, static lines and cantilever materials hoists (**working load limit** of 500 kg).

   Note: Height is limited only by the specifications of the manufacturer of the prefabricated components; however, 45 m is a commonly accepted maximum.
2. *Intermediate Scaffolding:* Basic Scaffolding plus tube-and-coupler scaffolds, cantilevered crane loading platforms, cantilevered and spurred scaffolds, barrow ramps and sloping platforms, screens, shutters and mast-climbing work platforms.
3. *Advanced Scaffolding:* Intermediate Scaffolding plus hung scaffolds, suspended scaffolds and the installation of all types of cantilevered hoists, including personnel and materials hoists.

All three scaffolding licences may only be obtained by persons over 18 years of age (this restriction applies to all High Risk Work Licences).

To obtain a licence in any of these categories, you must undertake training and instruction by a registered training organisation (RTO) with authority to run the appropriate units of competency. The licence itself is then issued by your relevant state or territory WHS/OHS authority.

* *Note:* Fall-arrest systems have limited application in scaffolding construction and in general should not be used (see 'Scaffold assembly: preliminaries' later in the chapter).

The necessity for a competency covering the erection and dismantling of restricted-height scaffolding (for which no licence is required) is found in the description of the duty of care in the Model WHS Act (and earlier state-applicable Acts and regulations). The duty of care requires employers (including the self-employed) to ensure that documented training is provided in all areas of work, and particularly in areas of high risk.

Restricted-height scaffolding: what sort of scaffold can I construct?

As stated in the introduction to this chapter, restricted-height scaffolding is limited to 4 m from the ground or support surface to the surface of the working platform (the highest surface upon which you can stand). Other than this, you can assemble all of the five scaffolding types outlined in this chapter.

Note: You *cannot* construct **tube-and-coupler scaffold** to *any* height. Figures 4.12 and 4.13 demonstrate when a licence would be required despite the actual height of the scaffold itself.

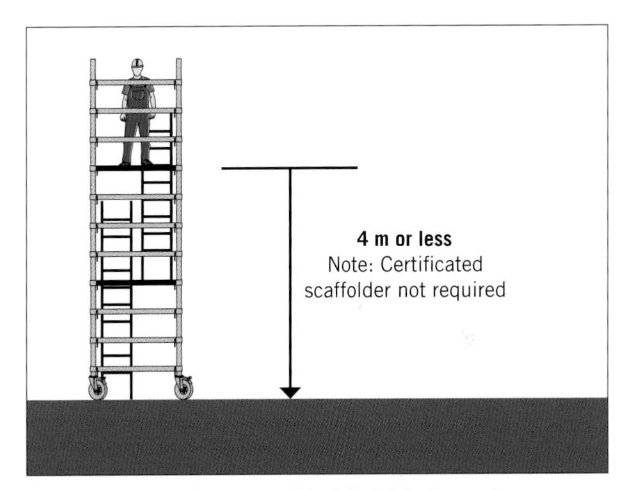

FIGURE 4.12 Maximum possible fall risk is 4 m or less: no licence required

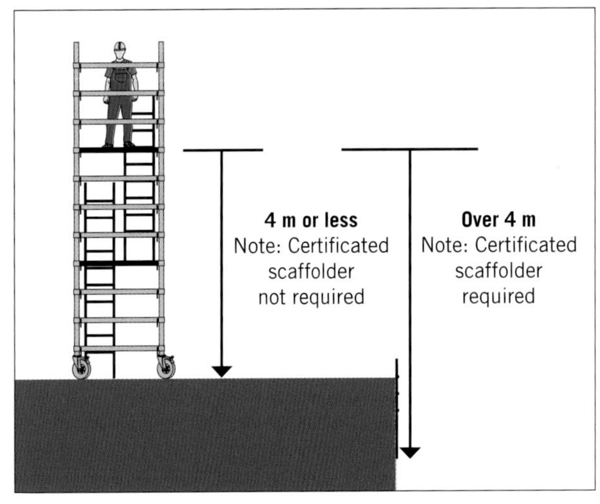

FIGURE 4.13 Maximum combined possible fall risk is determined as greater than 4 m: licence required

KEY POINTS

- Scaffolding is a level 2 control (from the hierarchy of controls).
- You are not permitted to construct tube-and-coupler scaffolding to any height without a licence.
- There are three licence levels you may apply for after further training: Basic, Intermediate and Advanced.
- You must be over 18 before you can obtain any High Risk Work Licence.

Scaffolding and the role of state and territory WHS/OHS authorities

Aside from issuing licences, your state or territory WHS/OHS authority has a significant role in ensuring that safe work practices and regulations are followed. Chapter 1 covers this topic in some detail; however, it is relevant to recap the basics here, and particularly how they pertain to scaffolding.

State and territory WHS/OHS authorities generally go by the name of 'SafeWork' or 'WorkSafe'. Their role is primarily to:

- administer the various national and state WHS/OHS Acts and regulations, including workers' compensation and injury management/return to work provisions
- provide advice and information on WHS/OHS issues
- monitor the industry for compliance with the various Acts and regulations
- investigate breaches or possible breaches of the Acts and regulations
- aid in the resolution of WHS/OHS disputes in the workplace
- audit workplace WHS/OHS systems and procedures.

In addition, WHS/OHS authorities can issue fines and initiate prosecution proceedings where non-compliance is strongly evident, has brought about injury, or could potentially bring about injury.

It is important to note that WHS/OHS authority inspectors may inspect a site without notice at any time.

With regard to scaffolding, WHS/OHS authorities require the following:

- the person responsible for constructing any scaffold above 4 m in height holds one of the three licences mentioned previously
- persons being trained for a qualification are under direct supervision by a licence holder at all times
- logbooks detailing each training session are kept by the trainee, with each entry being countersigned by the licence holder
- all scaffolding is constructed in accordance with the relevant Australian Standard (Generally the AS and AS/NSW 1576 suite: Scaffolding, Parts 1-6)
- a safe work method statement (SWMS) or its equivalent has been completed and signed off by all relevant workers and supervisors
- the SWMS is constantly reviewed in light of changing work conditions
- a site hazard assessment is conducted and countermeasures are put in place as required (see 'Planning the job: hazard assessment')
- the site hazard assessment is constantly reviewed in light of changing work conditions (see 'Planning the job: hazard assessment')
- the scaffold is regularly checked for serviceability by the licence holder and a written record of this is kept on-site (see 'Inspecting and maintaining the scaffold').

ON-SITE

NOT 'THE ENEMY'

It is important that both you and your employer have a good relationship with the regional or local WHS/OHS authority and their inspectors. They are there for your safety and to assist you in creating a safe, efficient workplace. *They are not 'the enemy'.* Treat them with the courtesy and respect they deserve for undertaking what is a difficult and, unfortunately, at times, horrifically stressful role (not dissimilar in many ways to that of the police). Follow their advice and your work will flow smoothly. Ignoring their advice is foolish at best, and could even prove deadly.

LEARNING TASK 4.1

1 **Circle 'True' or 'False'.**
The term 'scaffold' is limited only to those temporary structures that include access and working platforms.
True False

2 **At the conclusion of this unit of study you will be able to construct scaffolding to a height of not more than:**
a 2 m to the surface of the working platform
b 4 m to the handrail of the working platform
c 4 m to the surface of the working platform
d 6 m to the handrail of the working platform

Planning the job: hazard assessment

One of the key things that WHS/OHS inspectors will be looking for during an inspection is evidence that you have conducted a thorough hazard assessment prior to assembling a scaffold, and that ongoing assessments are made and documented once the scaffold is in place. The process is not dissimilar to that conducted for elevating work platforms (EWPs), as covered in Chapter 3. As such, you must plan for ongoing site inspections and ensure good communications with workers and managers from other trades. The recommended approach for assessing hazards is as follows:

- carry out a site inspection in consultation with a site supervisor
- discuss with other stakeholders
- ensure public and/or worker access
- determine and gain any required permits
- identify particular hazards.

This information will ultimately lead to the development of a SWMS.

We will now look at each of the above points in detail, before focusing on one hazard in particular: power lines.

Site inspection and consultation

Through this process, you will determine the layout of the site, access and egress routes, and the nature of the work needing to be done – that is, the purpose of the scaffold. This is your first step in determining the right scaffold and its location for the task at hand.

From your own observations, you should take note of:

- what the scaffold is to be used for (what work is to be done from it)
- the location and type of electrical sources, such as power lines, generators and switchboards
- support surface conditions (the type and nature of ground that the scaffold will stand on – includes bitumen or tarred surfaces, which may become soft with heat, or soils that may become boggy with water)
- trenches, drains and underground services (gas, electrical, water)
- areas of backfill
- drainage
- trees and obstructions
- roads and site traffic, and other vehicular movement around the proposed scaffold site
- wind loads (open or exposed area)
- general weather conditions (snow, ice, rain, heat)
- location of toxic, corrosive, explosive materials
- pedestrian traffic and other workers.

Most job sites will have someone who is effectively the site supervisor. (On small domestic sites, this may be simply the owner or builder.) From this person you should find out the following information:

- site policies and procedures (this may be as simple as signing in and out, for example)
- site induction procedures
- a copy of the plans of the works to be undertaken from the scaffold
- the job schedule for other trades
- support structures for surrounding works (including underneath the slab if you are working over a basement – you may need an engineer's assessment)
- particular site hazards (trenches, backfill, subfloor works, wind or other exposed condition issues)
- public and worker access (on domestic renovations, this may mean children).

Having consulted with the site supervisor, you will now have identified the other trades with whom you should also discuss your planned work. It is these tradespeople who can provide you with key insights into the nature of the work to be conducted from the scaffold (such as wrenching, twisting or other shock loads). They may also have a better appreciation of the material and personnel loads that may be applied at any one time.

Discussions with other stakeholders

This consultation is about how your work might influence or interrupt the work or interests of others. People who may have a 'stake' (whose lives or interests may be influenced by your activities), aside from other workers, may include:

- the local council
- owners and workers in other businesses
- pedestrians
- owners of neighbouring properties
- road and transport authorities
- police
- service suppliers (telecommunications, power, water, gas, etc.).

As with working with an EWP (see Chapter 3), it is better to have these discussions prior to proceeding, rather than waste time and money on hired scaffold while you sort it out mid-job.

With regard to other workers on the site, you must look carefully at where they are going to be and ensure that:

- they stop working while you construct the scaffold, and/or
- you stage the construction around their work needs
- you design the scaffold so that they can continue to access areas after it is built
- you design the scaffold so that workers and pedestrians below the structure are safe from falling tools, materials or equipment.

There may also be other considerations you should discuss with them, such as:

- dust
- noise
- interruption to power supplies
- limited or blocked access (for materials and/or workers)
- the need for additional personal protective equipment (PPE; such as hard hats).

On some sites, these discussions may need to be with collective worker organisations such as unions. Again, it is better to do this before, rather than after, you have set up and begun assembly. Stopping mid-job is expensive.

Public and/or worker access

With the above information in hand, you can then determine the best approach for ensuring public and/or worker access and egress around and/or on and off the scaffold.

The best solution to 'crowd control' (be it other workers or the public) while assembling a scaffold is the establishment of a 'no go' or exclusion area (see Figure 4.14). Means of establishing 'no go' areas include:

- witches hats (safety or marker cones)
- tape and mesh barriers
- non-climbable fencing
- warning signs
- barricades and barriers
- traffic control personnel
- trained spotters.

FIGURE 4.14 Establishing an exclusion or 'no go' area

Part of your discussions may include staging of your scaffold construction. In this way workers, including perhaps pedestrians from other places of work, may pass through the 'no go' area when the work is at a safe stage and there is no risk of falling objects. This may be for lunch periods, or specific times when the public, traffic, other workers or materials have a particular need to get through.

Determining and gaining required permits

While establishing a 'no go' area may be the best solution to keeping people and traffic out, you may need to gain a permit to be there yourself. Indeed, you may need to gain several permits. Authorities to which you may need to apply include many of the stakeholders mentioned previously:

- road and traffic authorities
- police
- local council
- electricity authority
- water board
- gas company
- telecommunications companies.

To determine to whom you should apply, the best approach is to assume that you will need a permit from one or all of the previously mentioned authorities if you:

- are near any of the above services
- block traffic or public access
- must build close to or over property other than the work site.

In other words, contact *them* before they contact *you*. They may then contact other stakeholders you have not thought of, paving the way for better relations all round.

Hazard identification

You are now in a position to identify and develop a response to possible site hazards. As mentioned previously, this may be by means of a SWMS. Effectively, this is a summary of your research, coupled with an action plan. Hazards you may have identified may be categorised under the following headings:

- Support surface:
  - backfill
  - excavations
  - basements
  - lower floor-level supports
  - underground services (gas, water, electricity, etc.)
  - voids, staircases, access holes and trenches.
- Scaffold erection area:
  - angled surfaces
  - rough or uneven terrain
  - unstable ground
  - ramps
  - obstructions
  - poor light
  - traffic (vehicular and pedestrian)
  - other tradespeople
  - wet or slippery surfaces
  - dangerous materials storage
  - power supplies (extension leads and other low-level supply lines such as gas or fuel).
- Overhead and other elevated hazards:
  - electrical supply lines (power lines)
  - climatic conditions (wind, rain, etc.)
  - maximum headroom
  - other works being undertaken at a higher level
  - nesting birds (plovers and magpies, in particular).

This brings us to a particularly potent hazard that is frequently underestimated and so needs some detailed comment: power lines.

Power lines

Power lines impose very particular hazards due to electricity's penchant for bridging or 'jumping' to the nearest conductor giving access to the ground (hence the electrical term 'to ground' or 'to earth').

Scaffolding components are generally formed of metal (steel or aluminium) and make excellent conductors. For this reason, the distance you must be from various line voltages is regulated to some extent by AS 4576, which requires that within Australia the clearance between scaffolds and any transmission line or apparatus be not less than:

- 4 m where any metal member is used
- 1.5 m where only non-conductive materials such as dry timber or plywood are used.

However, as is noted in the 'On-site' box on this page, there are significant differences between states and territories as to what is considered a safe distance from electrical sources. Because of this fact, it is important that you check with your local authorities to ensure compliance.

At times, you may need to install scaffold closer than the stated distances would allow. In such cases, you must again make contact with the relevant power supply and WHS/OHS authorities. Depending on the voltage of the line, the length of the line (due to swaying in the wind) and how close you are to it, you may need to do the following:

1. Install 'tiger tails' (split rubber insulators, named for their black-and-yellow striping: see Figure 4.15), which must extend a minimum of 5 m beyond either end of the scaffold. These are considered a visual guide only in some states and territories and so power must still be disconnected from these lines.

FIGURE 4.15 Tiger tails may be required should scaffolding need to be installed closer than stated distances from power lines

2. Install hoarding or shuttering of either the scaffold or the lines themselves (if they must pass through the scaffolding). This hoarding must be entirely of non-conductive material such as dry timber or ply.
3. Disconnect (de-energise) the power supply until all works are completed and the scaffolding is removed.

Note: All works (tiger tail installation, hoarding and disconnection) must be conducted and/or overseen by trained personnel from the power authority. Options 1 and 2 are rendered unsafe in wet weather conditions, and in such cases the scaffold must be evacuated.

ON-SITE

STATES/TERRITORIES DIFFER IN WHAT IS A 'SAFE DISTANCE'

Each state WHS/OHS authority has a different approach to specifying safe distances from power lines, and the distances vary markedly.

For example, Western Australia allows scaffold within 0.5 m of insulated 1000 V or less power lines. (This is the type of aerial power line connected to the fascia of a domestic home.) However, the same code states that if you are holding a scaffolding component that is 2.4 m long, then the closest you could go to the line would be 2.9 m (conductor length plus danger zone width). Tasmania, on the other hand, requires a minimum of 4.0 m clear distance from a line of the same voltage and type.

Due to these state and territory disparities, you must always make contact with your local power supply authority, and the relevant WHS/OHS authority, to determine the safe approach and construction distances.

Other conditions may also be imposed, such as:

- relocation or removal of power lines
- restricted work site entry
- control over work site activity by the power supply authority
- the use of trained safety observers, or spotters (see Chapter 3, 'The spotter: a critical role').

For distances relevant to your state or territory, go to:

- Australian Capital Territory: https://www.worksafe.act.gov.au/health-and-safety-portal/safety-alerts/overhead-electrical-risks
- New South Wales: http://www.safework.nsw.gov.au/__data/assets/pdf_file/0020/52832/Work-near-overhead-power-lines-code-of-practice.pdf
- Northern Territory: https://worksafe.nt.gov.au/forms-and-resources/bulletins/working-close-to-overhead-and-underground-power-lines,-gas-pipes,-and-other-infrastructure
- Queensland: https://www.worksafe.qld.gov.au/__data/assets/pdf_file/0017/20906/working-safely-near-powerlines_6317.pdf Also see the video: https://www.worksafe.qld.gov.au/resources/videos/films/electrical-exclusion-zones

- South Australia: https://www.sa.gov.au/__data/assets/pdf_file/0003/6969/160708-Working-safely-near-overhead-powerlines.pdf
- Tasmania: https://s0.whitepages.com.au/08a7535f-f806-4cc5-89c0-f3e3545fd0ea/tasnetworks-document.pdf
- Victoria: https://content.api.worksafe.vic.gov.au/sites/default/files/2018-06/ISBN-No-go-zones-for-overhead-electrical-power-lines-2004-07.pdf
- Western Australia: https://www.westernpower.com.au/media/2219/working-safely-around-the-western-power-network-factsheet.pdf

KEY POINTS

Planning the job is effectively five steps consolidated into a SWMS document:

1. site inspection (with site supervisor)
2. discussions with stakeholders
3. ensuring public and/or worker safe access
4. determining and gaining required permits
5. identifying particular hazards and necessary controls.

You must be over 18 before you can obtain any High Risk Work Licence.

Minimum distances from power lines are given in AS 4576, but these may be overridden by state or territory requirements.

To work closer than the minimum distance requires permission from the power supply authority and appropriate additional safety factors (tiger tails and/or hoarding).

COMPLETE WORKSHEET 1

Choosing the right scaffold

With the information you have gained through the hazard assessment process, you are now in a position to determine which of the various scaffolding options open to you is the most appropriate, and its design. At this point, you may also begin working on your SWMS, as one process informs the other (see 'Consolidation: creating the SWMS' later in the chapter).

In making your decision, reflect upon what you know of the site, the task and the various stakeholder needs from the following perspectives.

With regard to WHS/OHS

Can the scaffold:

- be constructed in accordance with the supplier's documentation or an otherwise approved design
- be constructed and dismantled without harm or risk to either scaffolders, other workers or site personnel, or the general public
- include the necessary and appropriate means of access and egress
- have safeguards to protect persons within the vicinity from hazards arising from its presence and use
- be used to carry out the necessary work without harm or risk to either workers on the scaffold, other workers or site personnel, or the general public?

With regard to location and context

Can the scaffold type and design allow for:

- the proximity of roads, public spaces and other adjoining properties
- other vehicular traffic, plant or machinery (including cranes and the like)
- the passage of other workers, non-workers and materials around the site
- the location of power lines and any hazardous substances
- the type and/or quality of foundation material, particularly point loads (be it ground, road surface or concrete slab)
- the shape of the building or structure it must surround or reach
- wind, rain, snow and other environmental impact factors?

With regard to purpose

Is the scaffold capable of:

- providing working platforms at the required heights
- providing working platforms of the required duty rating (heavy, intermediate or light – see 'Working platforms and duty ratings')
- carrying the required number of workers
- catering for the expected or possible live and dead loads that could be imposed (including people, materials, equipment, and the like)
- remaining stable for the duration of the tasks needing to be undertaken
- catering for works needing to be undertaken during and after its use
- being constructed and dismantled within the desired time frame?

We are nearing the point at which you would need to make your decision about which scaffold to use. However, before looking more closely at each of the five types you are likely to be called upon to construct, you need to come to grips with some key scaffolding elements and the regulations surrounding their construction and use.

LEARNING TASK 4.2

1 **Circle 'True' or 'False'.**
The shape of a building or structure has no bearing on the type of scaffold you choose.
True False

2 **The safe distance that a scaffold may be positioned from power lines is determined by:**
a your state or territory WHS/OHS authority
b the voltage of the power lines
c the type of scaffold being used
d both 'a' and 'b'

Working platforms and duty ratings

As stated earlier, the purpose of a scaffold is to create a safe and appropriate work platform at heights. Although in the next section we will begin our discussion of five different types of scaffold by which this may be achieved, the work platforms they support will generally fall into one of three categories. It is these categories, known as load or duty ratings, that determine the form a platform must take.

Duty ratings

Three categories of duty or load ratings are available for scaffolding platforms, these being: light, medium and heavy. It is important to have an understanding of these duty ratings, as well as working platform regulations, as the relative load limitations are inherent in the original manufacturer's design and intended end use.

Table 4.1 gives the size and load limits for the three main duty ratings as specified by AS/NZS 1576.1 and AS 4576.

TABLE 4.1 Loads and widths of working platforms

Duty rating	Maximum total load (persons, equipment and materials) per platform per bay	Minimum width (mm)	Minimum headroom above platform surface (m)
Light*	225 kg	450	1.85**
Medium	450 kg	675	
Heavy	675 kg	900	

**Note:* Materials may not be stored on light-duty work platforms that are of the minimum allowable width.

***Note:* Where it is structurally impossible to maintain this headroom in a localised area, it may be reduced to a minimum of 1.72 m (e.g. at transoms, lapped planks, plan bracing, ties and the like).

Other duty ratings

AS/NZS 1576.1 actually describes five duty ratings – the three in Table 4.1 plus closed and special-duty platforms.

- *Closed platforms:* These platforms are blocked off from access or storage of materials and are not used. A **closed platform** can be opened and used as a working platform when required and so must be constructed to the same specifications as a heavy-duty platform.
- *Special-duty platforms:* This applies to platforms that are of a dimension that does not fit the other ratings (light, medium, heavy). What is known as a **birdcage scaffold** (see Figure 4.7) is often used to produce working or storage platforms that are of greater width than normal. Calculation of special-duty loads is based upon uniformly distributed loads and so a maximum pressure (kPa) per square metre (m^2), rather than a kg or kN (kilonewton).

Duty action

Duty action is the load applied to a working platform within a bay. It includes:

- the weight of persons
- the weight of materials and any debris
- the weight of tools and equipment
- impact forces.

Note: The weight of a single person is taken as not less than 1 kN or approximately 102 kg as concentrated load.

As Tables 4.1 and 4.2 demonstrate, the size and load-carrying capacity of working platforms is carefully regulated.

TABLE 4.2 Duty action loads of working platforms

Duty action	Maximum total load per platform per bay	Maximum concentrated or point load (as portion of total load)
Light	2.2 kN (~225 kg)	1.2 kN (~120 kg)
Medium	4.4 kN (~450 kg)	1.5k N (~150 kg)
Heavy	6.6 kN (~675 kg)	2.0 kN (~200 kg)
Special	Largest intended load (not less than 1 kPa)	
Loading platforms	Largest intended load × 1.25 to allow for impact (not less than 5 kPa)	
Bay extension platforms	Equal to light-duty platforms	

Other working platform regulations

There are several additional factors about working platforms that you should be aware of before we proceed further.

Figure 4.16 shows a typical work platform on a modular scaffold and its component parts. It will be helpful at this point if you spend a moment coming to grips with some common scaffolding terminology. (A full glossary of terms can be found at the end of the book.)

- *Guardrail:* A structural member to prevent persons from falling off any platform, walkway, stairway or landing.

FIGURE 4.16 Heavy-duty working platform

- *Ledger:* A horizontal structural member longitudinally spanning (running the long way) between two standards (generally) on a scaffold.
- *Midrail:* A member fixed parallel to and above a platform, between the guardrail and the platform.
- *Putlog:* A horizontal structural member, spanning between ledgers and standards or between these and an adjacent wall. Its purpose is to support the surface of the working platform – that is, you 'put the logs on it' to make up the platform. Logs have been replaced by planks, but the name has stuck.
- *Scaffold plank:* A decking component, other than a prefabricated platform, that is able to be used in the construction of a platform supported by a scaffold. Its minimum width is 220 mm (nominally 225 mm). It is supported by putlogs. The minimum working platform (light duty) is two planks wide, or 450 mm.
- *Standard:* A vertical structural member that transmits a load to a **supporting structure** or ground level: generally, this is by means of an **adjustable base plate** or screw jack. Except for planks and toe boards, the other components listed are directly connected (at least at one end) to a standard.
- *Toe board:* A scaffold plank or purpose-designed component fixed on its side at the edge of platforms to stop tools, materials and feet slipping off the side or end of a platform.
- *Transom:* Similar to a putlog in being a horizontal structural member transversely spanning (running the short way) between the standards of an **independent scaffold**. Transoms may serve as guardrails or midrails, or simply as spreaders. In some states, the terms 'putlog' and 'transom' have become interchangeable (or 'putlog' not used at all).

Note: Midrails and guardrails are generally the same components that are used for ledgers, **transoms** or **putlogs**. It is their location that dictates what role (and, hence, name) they will hold.

Constructing the working platform

A working platform should be constructed from purpose-designed, prefabricated components or scaffolding planks wherever possible. Depending on the intended duty (light, medium or heavy), the required width of the platform will change (see Table 4.1).

Irrespective of the duty rating of the platform, a clear passage of 450 mm width must be maintained at all times to allow passage for people, tools and materials. This is one reason (the other being weight) why materials and equipment cannot be stored on a light-duty platform as it is only 450 mm wide.

Working platform surfaces must also comply with the following:

- Surfaces must be horizontal wherever possible, or have a slope not greater than 3 degrees in any direction unless specifically designed and engineered for that purpose. The maximum gradient for specially designed platforms is then 1:8 (rise not more than 1 m over every 8 m, or 7 degrees).
- They must have planks or decking of uniform thickness and a non-slip surface.
- Planks must be secured to ensure they cannot be displaced.
- There should be no gaps or trip hazards.
- Planks should be butted end to end, not lapped, except at **returns**, curved surfaces or unusual profiles where no alternative is available.

Edge protection and guardrails

Edge protection is the installation of barriers to prevent people, tools and/or materials from falling from a platform. Generally, this means toe boards, midrails and guardrails, but may include wire mesh and/or hoarding (fully enclosing materials such as plywood).

AS/NZS 1576.1 and AS 4576 require that whenever a person can fall 2 m or more from a scaffold or working platform, edge protection *must* be installed. The standards also state that, irrespective of the fall height, if the work being undertaken on the platform is such that you might easily fall off the platform, then edge protection shall be installed anyway (e.g. doing work overhead, or using PPE such as goggles or welding helmets).

Appropriate edge protection where a fall may be 2 m or greater is described below.

The 'open' (non-work face) side of the platform

For the 'open' (non-work face) side of the platform (see Figure 4.17), appropriate safety measures are outlined below.

FIGURE 4.17 Edge protection requirements: open side

- Guardrails:
  - must be installed between 0.9 m and 1.1 m above the platform surface
  - must not be more than 3 m between supports (such as standards or posts)
  - must be made of scaffolding tube, or be a purpose-designed component, or be from long-grained hardwood, Oregon or LVL (laminated veneered lumber) with a nominal sectional size of 100 mm × 50 mm
  - must be located not more than 100 mm outside a line taken vertically from the edge of the platform.
- Toe boards:
  - should be a scaffolding plank on edge, or its equivalent in rigidity and strength
  - must be securely fixed
  - must have a minimum height of 150 mm above the surface of the platform
  - must not have a gap greater than 10 mm between it and the surface and/or edge of the platform.
- Midrails:
  - must be positioned midway between the guard-rail and the toe board; otherwise must have the same requirements as guardrails
  - panels and brick guards – these are designed to reduce the likelihood of materials, tools and other equipment falling from the platform. They may form part of the edge protection, with the following provisos:
    - if taking the place of a guardrail and midrail, they must be at least as strong as those rails would be normally
    - if made of steel mesh, the mesh must be 4 mm in diameter and have apertures not greater than 50 mm × 50 mm
    - they must include a **kickplate** or toe board to 150 mm above the platform surface
    - if made of plywood, it must be at least 17 mm thick.
- Access openings:
  - in guardrails to work platforms, must be fitted with self-closing gates that open on to the platform (not outwards)
  - when not on work platforms, may be closed with chain barriers
  - shall have no sharp edges or elements that may cut persons entering the platform
  - ladder or stair openings shall not be on or close to working platforms
  - access hatches shall not require a person to hold them open while entering or exiting, and must be positioned away from working platforms.

    Note: Wire or fibre rope must not be used as either guardrails or access closures.

For the work face side of the platform

Generally, one side of your work platform will be facing the building being constructed (see Figure 4.18). In such cases, no edge protection would be needed if the work face or building:

- is less than 225 mm from the platform edge*
- extends at least 900 mm above the top surface of the platform
- has strength and rigidity at least equal to that of a guardrail
- can perform the function of a guardrail and midrail in all other respects.

Access and egress to working platforms

All working platforms need a means of access and exit (or egress). How you decide to access a platform will depend upon the information you received during your consultations on hazards and the work to be done from the scaffold – that is, how many workers, what tools and materials need to be taken up, and by what means materials may be transported (wheelbarrow, for example).

---

* *Note:* A catch platform (refer to Figure 4.11) may be required where the distance is close to the 225 mm limit. Catch platforms help prevent tools and materials falling between the face of the work and the scaffold.

FIGURE 4.18 Distance from work face

Accessing or exiting a working platform by means of climbing the scaffold is prohibited.

The means of access and egress is generally by either:

- existing floor levels
- permanent platforms
- stairs (temporary or permanent) (see Figure 4.19)
- personnel hoists (see Figure 4.20)*
- access ramps**
- ladders.

Access ways must be constructed to the same requirements as heavy-duty working platforms. However, they may be:

- reduced to 675 mm when used for transporting materials (assuming materials and any transportation, such as wheelbarrows, will fit)
- reduced to 450 mm when used by workers carrying hand-held tools only
- sloped along their length to a maximum of 20 degrees, or 1:3.***

Scaffold planks

The surface of a working platform is generally produced by timber or metal planks. But not just *any* plank. Scaffolding planks (Figure 4.21) are carefully designed. Their limits are dictated by AS/NZS 1577 Scaffolding planks, and their uses by the AS series AS/NZS 1576 and AS 4576. These standards stipulate the need for a slip-resistant surface, that they cannot slide easily, as

* *Note:* If using a personnel hoist as a means of access, an alternative means of egress must be made available, such as ladders or stairs. Hoists may only be installed by holders of an Advanced Scaffolding licence.

** *Note:* Access ramps may only be installed by holders of an Intermediate (or higher) Scaffolding licence.

*** *Note:* When sloped at greater than 1:8 (7 degrees), access ramps must be cleated to prevent slipping. Cleats are to be 25 mm high and 50 mm wide, extend the full width of the platform/ramp, and be fixed at 450 mm centres. A 100 mm section may be removed from the middle of each cleat to leave a clear passage for wheelbarrows if required.

Source: Layher Pty Ltd

FIGURE 4.19 Scaffold stairs

Source: Layher Pty Ltd

FIGURE 4.20 Personnel hoist

FIGURE 4.21 An example of a timber plank with branding

well as issues such as rigidity, twist, and the like. Table 4.3 outlines the key requirements for scaffolding planks of which you should be aware.

All scaffolding planks must be permanently marked or branded,* offering the following information:

- manufacturer's name or identification
- the Australian Standard with which it complies (e.g. AS/NZS 1577)
- working load limit in kilogram
- if random length planks – the allowable span in metres
- if timber – 'V' if visually graded, 'M' if machine graded.

Do not use scaffolding planks with the following defects or damage:

- timber planks:
  - with breaks, splits or cracks
  - with damaged or broken metal end straps
  - that have lost or damaged end cleats (metal, plastic or timber)
  - that are excessively worn down (having lost 10% or more of their thickness)
  - with any evidence of rot, decay, or borer or termite damage
- metal planks:
  - with missing end caps
  - that have been crushed or bent (even if it has been straightened)
  - with rust or corrosion
  - with welds that are cracked or broken.

Now that you have a clear image of the types of platforms you are required to support, it's time to have a look at the basic design and general assembly procedures of the various types of scaffolds under review. But first, we will briefly consider the preliminaries of assembly, which apply to all types.

KEY POINTS

Working platforms are rated for their intended duty. AS/NZS 1576.1 specifies maximum **duty ratings** and **duty action loads**. In each case, **maximum point loads** are also given.

Working platforms are rated as:

- light (225 kg)
- medium (450 kg)
- heavy (675 kg).

All working platforms must:

- allow 450 mm clear passage
- have edge protection (guardrails, midrails and toe boards) where a fall of 2 m or greater is possible
- appropriate means of access (e.g. stairs, ladders, ramps or the like)
- be a minimum of 450 mm wide (or two planks of 225 mm).

COMPLETE WORKSHEET 2 AND 5

TABLE 4.3 Scaffolding plank requirements

Plank material	Width (mm)	Can be painted?	Thickness (mm)	End treatment	Minimum point load capacity
Timber: softwood	220–225	No	38	Metal strapped or through-cleated with timber, metal or plastic inserts	2 kN (200 kg)
Timber: hardwood	220–225	No	32*	Metal strapped or through-cleated with timber, metal or plastic inserts	2 kN (200 kg)
Timber: LVL	220–225	No	35	Splinter free	2 kN (200 kg)
Metal: aluminium	220–225	Yes (anti-corrosive)	NA**	Folded and no bursor sharps	2 kN (200 kg)
Metal: steel	220–225	Yes (anti-corrosive)	NA**	Folded and no bursor sharps	2 kN (200 kg)

**Note:* This is dependent upon the strength group rating of the timber. See AS/NZS 1577, clause 3.5.1.2, and AS 1684 Residential timber-framed construction for details on specific timbers. For example, a spotted gum plank can be 32 mm, whereas a mountain gum plank would need to be 38 mm thick.

***Note:* No thickness specified. Must comply with relevant Australian Standards (steel/aluminium) regarding tensile strength and elongation. Commonly 40–50 mm thick.

**Note:* For timber planks, branding is to be not less than 25 mm high and not more than 1.8 m apart.

Scaffold assembly: preliminaries

Despite the many and varied types of scaffolds available, there are some general preliminaries applicable to all. These include WHS/OHS, site preparation and some common tools.

WHS/OHS

Erecting scaffolding has a number of basic and inherent risks associated with working at heights and working with long metal and/or timber components overhead. Despite Australian Standards requiring components to be 'sharps' free, the nature of the components and their use means that some burrs and/or splinters may be present. In addition, depending upon the weather conditions and level of exposure, components can become extremely hot, cold and/or slippery. Generally, you are also not working alone. All this means that the following PPE is required when assembling or dismantling scaffolding:

- *hard hats* complying with AS/NZS 1801 Occupational protective helmets. Hard hats must:
  - have chin straps to prevent them falling when you bend over
  - be tested or replaced every two years
  - not be dropped or damaged
  - not be marked with felt pens, stickers or paints
- *safety footwear:* Boots must be steel capped, flexible enough to be non-slip on uneven surfaces, yet offering good ankle support
- *high visibility (HiVis) vests or tops*
- *rigger's gloves:* Protection from cuts, bruising and abrasions, as well as from heat and cold
- *eye protection:* To protect against metal shards coming from wedges and pins when being driven
- *ear protection:* Reusable or disposable plugs that still allow good levels of human speech to be heard, yet reduce the loud metal-on-metal impact noises which frequently exceed daily maximum decibel duration levels
- *tool belt:* Allowing you to climb using both hands
- *clothing:* Long-sleeved cotton shirts, and cotton trousers or overalls, offering protection against UV as well as cuts and abrasions.

Manual handling

In addition to the PPE, there is some fundamental skill and knowledge required regarding manual handling. These skills will have been covered in the prerequisite competency, CPCCWHS2001: Apply WHS requirements, policies and procedures in the construction industry. To recap the basics:

- Always bend your knees and lift with your legs, not your back.
- If you cannot lift an object in the above manner, then it is too heavy for you.
- Never attempt to lift anything over 20 kg by yourself.
- Do not over-extend or overreach when pulling, lifting, or passing materials or tools.
- Never climb a ladder while holding tools or equipment.
- Maintain a clear work area.
- Pass, put or place scaffolding components. Never throw them.

Remember, scaffolding is a team exercise: there are individual responsibilities, but the team must be able to rely on you, and you on the team.

ON-SITE

A NOTE ON INDIVIDUAL FALL-ARREST SYSTEMS (IFAS)

Individual fall-arrest systems (IFAS) and travel restraint systems have very limited application in general or basic scaffolding (see Chapters 1 and 3 for details of these systems). Such systems are not only a hindrance, but also impose a significant hazard that may lead to a fall. In addition, should a fall occur, the wearer is not protected from injury as they may strike the scaffolding multiple times before the lanyard deploys fully, only to be swung back and strike the scaffold again (see Figure 4.22). For these reasons, it is generally considered best practice that IFAS should not be used. Such systems are only recommended during the construction of suspended scaffold, where there is nothing beneath the wearer but the proverbial long drop. In such cases, advanced scaffolding training is required.

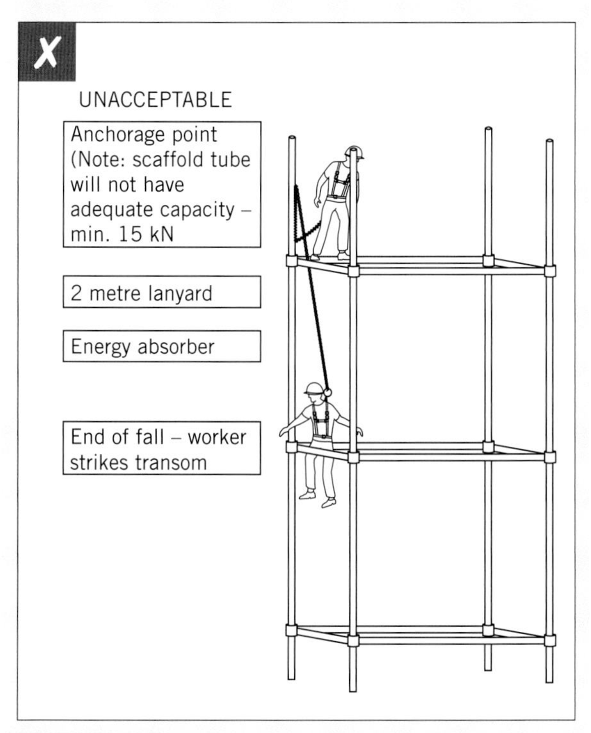

FIGURE 4.22 IFAS can impose a hazard when used in scaffolding structures

Site preparation

This is effectively the practical application of the outcomes from your hazard assessment conducted earlier. As a minimum, you will need to do the following:

- make contact with the site supervisor and ensure that all relevant workers are aware of your intended actions
- establish appropriate areas for the location of scaffolding components. Areas should:
  - be close to the scaffold construction area
  - allow differing components (**ledgers**, **braces**, **standards**, etc.) to be placed in separate locations to ensure ease of access
  - ensure that emergency access to the scaffold location is maintained
- ensure that the area for the scaffold is clear of debris, materials and equipment
- set up a 'no go' area, or otherwise control access to the assembly area
- check that all hazards have indeed been identified and appropriate controls put in place. Pay particular attention to power lines and the state of the foundations or support surface
- check that all permits required have been obtained
- check that SWMS document is complete, and the agreed procedures to be followed
- check that all those working on the scaffold are suitably trained and qualified
- check that all appropriate PPE is worn.

Upon delivery of scaffold materials

Once the scaffold components have been delivered, and preferably during unloading, you should make an immediate check of the following:

- you have the correct number and types of components
- the components have been located on-site appropriate to the manual handling needs determined previously (see 'Manual handling' earlier)
- any damaged, defective or mismatched components are isolated and then removed from site as soon as possible
- emergency access for essential services has not been blocked or hindered
- the supplier has provided the following information:
  - instructions for erection, maintenance, dismantling, transport and storage
  - a guide for safe work practices, including guidance on stability of the scaffold
  - guidance on the correct type of scaffolding **couplers** for connecting ties and other tube elements
  - duty rating of scaffold and maximum working platform capacities
  - maximum number and array of working platforms
  - maximum permissible height of scaffold. (You, of course, are limited to constructing it to 4 m, despite its capacity to go further.)

During scaffold construction

Again, some basic principles apply to all scaffold types as they are being erected:

- make sure you are following your SWMS procedure
- ensure the correct PPE is being worn
- take note of any changed site vehicular and/or pedestrian/worker traffic paths and compensate accordingly (by installing other barriers, signage or spotters)
- monitor scaffold components for defects, and mismatched materials, such as types of connections and/or differing manufacturer brands (see 'Connecting modular components: do not mix-and-match' later in the chapter)
- ensure that manual handling and lifting procedures are appropriate
- if the scaffold is incomplete and you have to leave the site, ensure the following:
  - effective barriers are installed to block access to work platforms
  - all barriers include signs warning that the scaffold is incomplete and entry is prohibited
  - access to the scaffold at support surface level is fenced off, with warning signs as above
  - the scaffold has reached a secure and stable stage of erection capable of withstanding likely imposed loads such as wind, rain, snow and the forthcoming continuing assembly.

The basic tools

Not all of these tools will be required for every type of scaffold we will be discussing; however, they are generally the most common ones required.

Remember, before using any plant, tools or equipment it is important that they are checked for any signs of damage and are fit for purpose. If any damage or faults are identified this must be reported to the supervisor immediately and followed up in writing. The piece of equipment should also be taken out of circulation and tagged out.

- *Scaffolding belt* (Figure 4.23): This belt has what is known as 'frogs' into which, or on to which, the various other tools may be stowed (safely and securely carried as you move and work). The belt should be made of leather or webbing of similar strength.

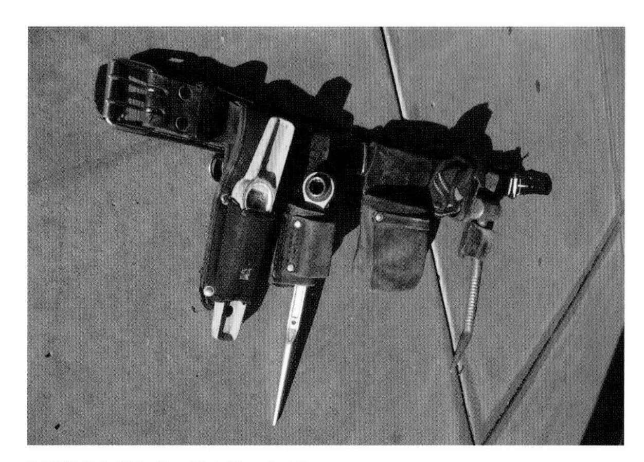

FIGURE 4.23 Scaffolding belt

- **Scaffold spanner** or **key** (Figure 4.24): These are used to tighten and loosen the nuts on couplings. The handle should be between 200 and 250 mm in length to ensure ease of disassembly while not allowing excessive tightening. If the socket of the tool becomes worn or damaged, the tool should be replaced.

FIGURE 4.24 Scaffolding key

- *Shifting spanner* (also known as an 'adjustable wrench' or a 'shifter' – Figure 4.25): 'Shifters' should only be used when it is not possible to get ready access with a scaffold key. The jaws of shifters tend to open a small amount when loaded, and this can damage the nuts on couplings and/or cause the tool to release rapidly, causing injury. Care should be taken, and their use limited. Handle length is as for a scaffold key: 200–250 mm.

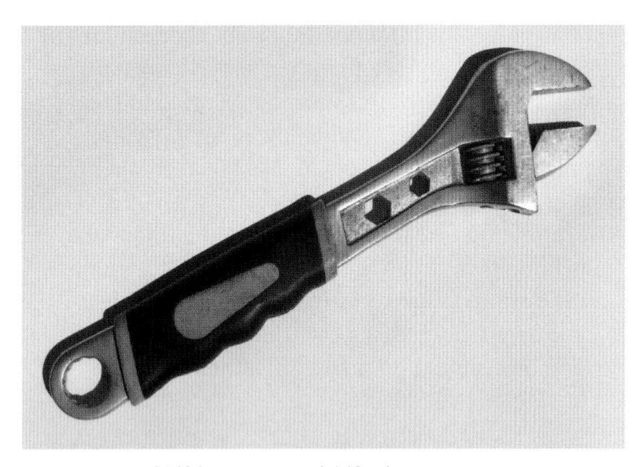

FIGURE 4.25 Shifting spanner (shifter)

- **Podger hammer** (Figure 4.26): This is a steel mallet or hammer with a bent handle ending in a spike or 'podger'. The mallet end is used for fixing wedges, pins or other locking systems used on scaffolds. The podger is used for levering wedges and pins as required. (Scaffold keys can also be shaped so as to have a podger end.)
- *Spirit levels*: Generally only a small magnetic level is required (150–250 mm in length). The short length allows them to be stowed on a belt; the magnetic strip on the edge allows the level to stay in place on a component while you adjust the scaffold accordingly. Levels should be checked on a regular basis.

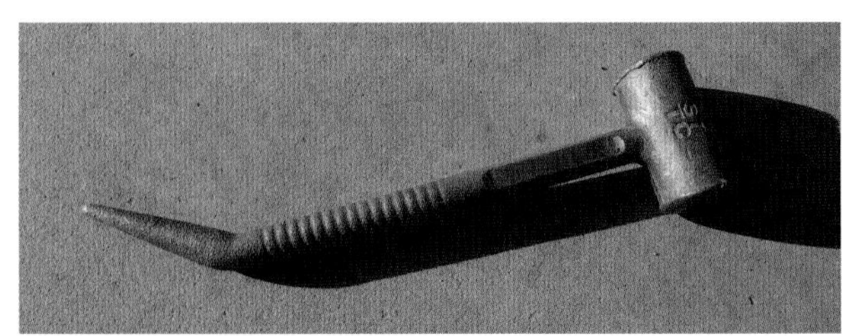

FIGURE 4.26 Podger hammer

- *Tape measures*: 5 metre and 8 m tapes are usual for scaffolders. This allows you to check diagonals on setting up. It is also a size easily carried in the front frog of your tool belt.

Dismantling of scaffolding

The dismantling and storage of scaffolding shall be discussed with regard to each of the various scaffolding types, which means looking at how to assemble the various scaffolds themselves.

We shall now begin our discussion of the five selected scaffolding types with the most common type used in Australia: modular scaffold.

LEARNING TASK 4.3

1 **Circle 'True' or 'False'.**
All working platforms must have a minimum of 450 mm clear passageway for people to pass by materials or co-workers.
True False

2 **The maximum total load on a heavy-duty scaffold platform is:**
a 2.2 kN or 220 kg
b 4.4 kN or 440 kg
c 6.6 kN or 660 kg
d 8.8 kN or 880 kg

Modular scaffold

For working to heights approaching your 4 m limit, this is one of the most common forms of scaffold that you will come across. If the working platform you require is heavy duty then, with the exception of unit frame scaffolding, it is one of the few that will achieve it. Indeed, **modular scaffold** is commonly used to heights of 30 m and more on multi-storey constructions.

Figures 4.17 and 4.18 offer you the main components of a working platform for this form of scaffold. Figure 4.27 expands on that information by showing the component parts of one **lift** of one bay of modular scaffold. We will start by coming to grips with these two terms first.

- *Lift*: This is the vertical distance from the base to the lowest putlog at which a platform could be constructed. It is also the vertical distance between adjacent ledgers at which a platform could be constructed (generally, 2 m – see Figure 4.27).

- *Bay:* The space enclosed by four adjacent standards arrayed in a cage with a rectangular or square floor pattern (see Figure 4.27). The same term applies to the equivalent space in a **single-pole scaffold**.

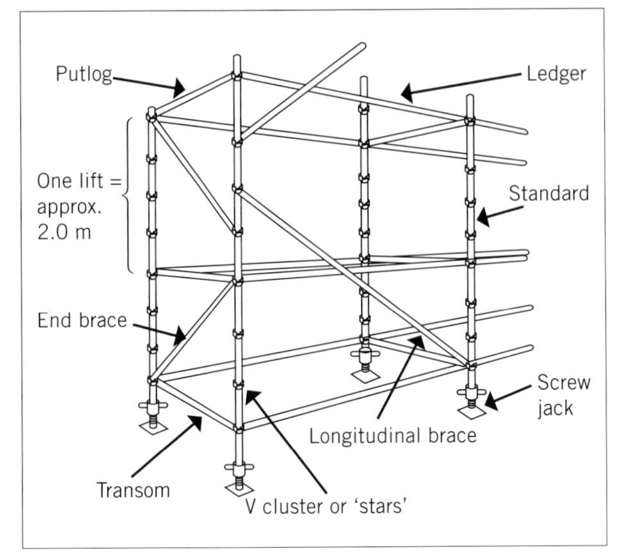

FIGURE 4.27 Components of one lift of one bay of modular scaffolding

The components of modular scaffold are generally made from steel, steel alloy or aluminium. The standard dimensions* of common scaffold tube are set out in Table 4.4.

TABLE 4.4 Common tube dimensions for modular scaffold

Tube material	External diameter (mm)	Wall thickness (mm)
Steel	48.3	4
Aluminium	48.4	4.4–4.5

Connecting modular components: do not mix-and-match

All the components of modular scaffold clip together using patented locking or wedging systems. Figure 4.28 shows three of the more commonly available types.

It is critical that you learn to identify between systems, and *only use components of the one system in any given scaffold assembly.* This is because some systems do 'sort of' fit together (with a bit of persuasion), but this will lead to stressed joints and possibly catastrophic failure.

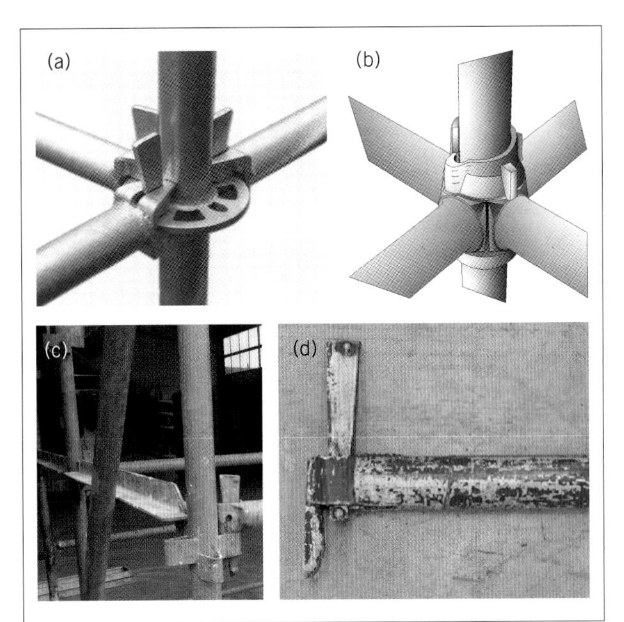

FIGURE 4.28 Common component connection systems: (a) ringlock (b) cuplock (c) kwik stage (d) pin and wedge, kwik ledger

It is also important that you *only use components of the one brand or manufacturer.* If a supplier provides you with components of more than one brand, then, as per AS 4576, they must also supply you (or your supervisor) with a four-part document, signed by the owner/manager/director, guaranteeing that:

- components are compatible in size and strength
- they have compatible deflection
- they have compatible jointing systems
- the mix of components does not lessen strength, stability, rigidity, or the ultimate suitability of the scaffold for the task for which it is required.

Do not mix aluminium components with steel components. This is because aluminium and steel have very different deflection, strength and weight characteristics. For example, aluminium standards are designed for the load of the working platform limits and the scaffold weight based upon the weight of aluminium. Mix in steel and you may overload the standards. On the other hand, aluminium ledgers will flex more than steel. Mix these with steel ledgers and standards and you will impose greater point loads on the steel components, because one or two aluminium ones are flexing when they should not and are effectively not doing their job. Either scenario can lead, and has led, to collapse.

Generic assembly procedure

As stated earlier, scaffolding is a team task, and so scaffold should be assembled by a minimum of two people working together at all times. In following this procedure, ensure you do the following:

- maintain good, clear communications between the members of the scaffolding team

* *Note:* AS/NZS 1576.1 and AS 1576.3 allow almost any diameter and wall thickness, with some provisions to ensure that sizes are discernible from those more commonly in use. This is provided that the tube complies with the following:

- steel: AS/NZS 1163 Cold-formed structural steel hollow sections
- aluminium: Wrought to comply with AS/NZS 1664 Rules for the use of aluminium in structures; extrusions with AS/NZS 1866 Aluminium and aluminium alloys – Extruded rod, bar, solid and hollow shapes; and sheet and plate with AS/NZS 1734 Aluminium and aluminium alloys – Flat sheet, coiled sheet and plate
- performs to section 4 of AS 1576.3 with regard to load bearing and deflection.

- never turn your back on a partially supported standard unless it is clear that your partner has full control and is aware of what your next move is going to be
- keep tools in your belt when not using them. Never hang hammers on scaffold components, or leave any tool lying on planking.

Packing-screw jacks and pigsties

Screw jacks (Figure 4.29) or adjustable base plates (or legs) are the footings of your scaffold and so must be treated with respect. The purpose of the screw jack is to allow you to adjust the height of your standards so that the first row of ledgers (and hence the first lift) are level. Most jacks are approximately 700 mm in height with a **base plate** of at least 150 mm × 150 mm. The important thing to remember with screw jacks is that at least 150 mm of the screw or pin is allowed to extend into the standard that sits upon it. Any less, and there is the danger of the screw jack and standard buckling where the two components join.

FIGURE 4.29 Screw jack

The screw jacks themselves often have inadequately sized base plates to prevent them from sinking into the ground or otherwise putting too great a point load on to a supporting surface (such as pedestrian pavements, timber floors and the like). To reduce this point load, we use **sole plates**. Where necessary, we use multiple sole plates stacked one upon the other in alternating directions (Figure 4.30). This is known as pigsty packing.

Sole plates are generally of timber, and should at least be of similar sectional size to scaffolding planks (220 mm × 32 mm). In length, it is preferable that the one sole plate be able to bridge between at least two standards or screw jacks, though this is not always practicable due to the slope of the ground. When used singly, the minimum length is 500 mm.

FIGURE 4.30 Sole plates reduce point load – if more than one is used, they should be laid pigsty fashion as shown

When laying pigsty packing, be sure to begin by levelling the earth (with a shovel or by simply scratching at the ground with the end of the timber) under each of the first timbers.

Options for installing edge protection

Before starting your assembly, you must adopt a safe approach or procedure for positioning your guardrails and midrails. This procedure will be agreed to by all those working to assemble the scaffold and will be written into your SWMS.

There are effectively three options available that allow you to install these components without exposing yourself to a fall of 2 m or greater. These are discussed in order of preference below.

Option 1: Portable erection platforms

Various forms of these are available, from types similar to (but significantly better than) fold-out step ladders (Figure 4.31), to fixed-step platforms (Figure 4.32). In either case, the platform must comply with the AS series AS 1892 Portable ladders. This means the platform must:

- have a standing area of a minimum 350 mm wide × 300 mm deep
- have stiles not less than 350 mm apart
- be load rated to 120 kg (minimum)
- have non-slip feet
- either be self-supporting or able to be easily secured to the scaffold (by hooking to the lower guardrail, for example).

  For other stipulations, see AS 1892.1.
- *Pros:* Simple, quick and light to move; good folding ones clip easily to the guardrails.
- *Cons:* Non-folding ones can be awkward to move around and block access. Need to be removed from scaffold or otherwise safely stored once erection is complete. May need two for longer components (ledgers) to be positioned.

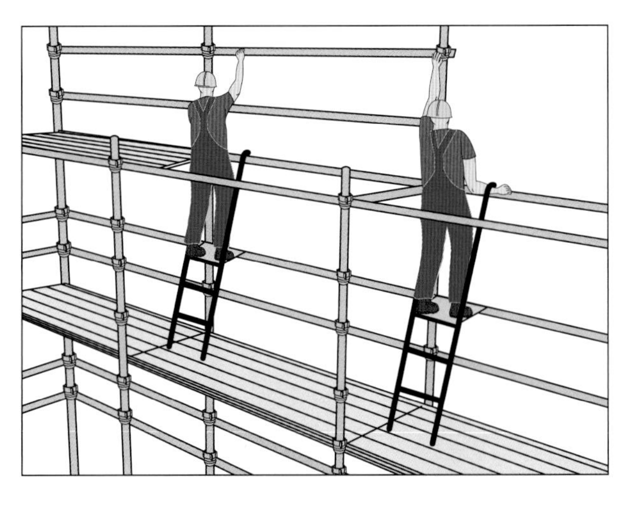

FIGURE 4.31 Fold-out step ladder fitted to scaffold

FIGURE 4.32 Fixed-step platform on scaffold

Option 2: Temporary working platforms

This involves installing additional putlogs 1 m below the target platform, and then laying in one or more scaffolding planks. You can now reach up and install the midrail, followed by the guardrail for that bay. The temporary platform is then removed and relocated to the next bay and the process repeated (see Figure 4.33).

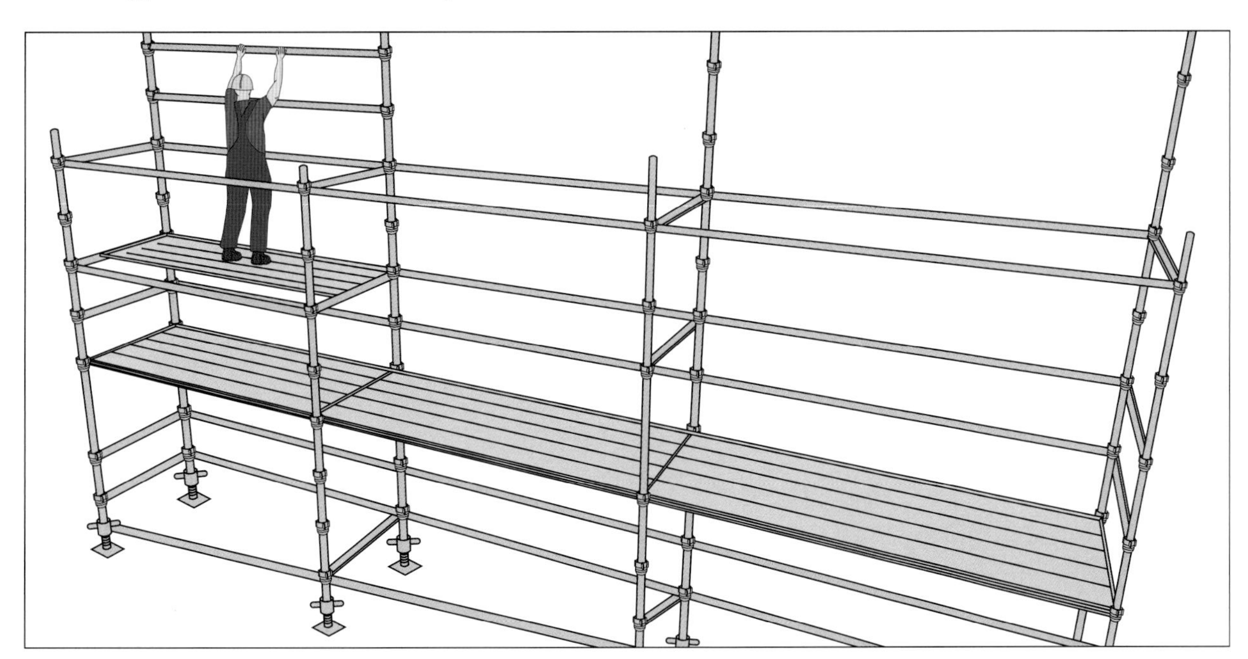

FIGURE 4.33 Installing temporary platform to fit midrails and guardrails

- *Pros:* Gives a full bay-length platform to work from. Components ultimately become part of the scaffold and do not need alternative storage.
- *Cons:* Labour-intensive. Putlogs need to be constantly repositioned to create the platform in the next bay. (Putlogs cannot stay in place, as they would block passage on the platform.)

Option 3: Temporary guardrails

This is a welded metal component made up of two or more uprights, a guardrail and a midrail (Figure 4.34). Temporary guardrails hook on to the ledger of the target platform, and the guardrail of the lower platform, as shown.

- *Pros:* Provides a secure and simple guardrail.
- *Cons:* Needs two people to position. Must be installed on the outside of the scaffold. Can be a falling object risk even if a safety chain is fitted (can swing into the lower platform). If not fitted correctly, can give a sense of security which will be found to be false when it collapses. Must be removed from scaffold or otherwise safely stored.

FIGURE 4.34 Temporary edge protection

HOW TO

PROCEDURE FOR ASSEMBLING MODULAR SCAFFOLD

What follows is a sequence of steps that, if followed correctly, provides you with a safe procedure for assembling modular scaffold to any height (remembering that you are limited after this training to a maximum of 4 m).

Step 1	Locate position of first screw jacks, level the surface, and establish sole plates where necessary. ***Guidance note 1:*** On sloping ground, it is best to start at the highest point and work down. Screw jacks at the high end should be wound down, leaving approximately 50 mm of adjustment for final levelling. ***Guidance note 2:*** Use a transom (putlog) and ledger as your means of locating the jacks, and leave them in position so they can be lifted into place later.	

>>

>>

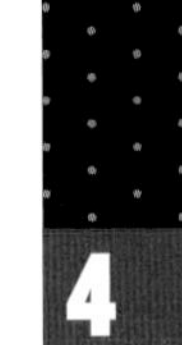

Step 2	Bring up two standards. Position the first standard on a jack and have one person hold this while you put the second standard in place. One person can now hold two standards while the other positions the first transom. ***Guidance note 1:*** Be sure to rotate the standard so that the highest wedge or pin receiver (known as 'stars', or 'V' clusters) is aligned with the ledger, and the lower with the transom. This allows the ledgers to sit slightly higher than the putlogs and so lock the planks in on working platforms. (This does not apply to cuplock systems.) ***Guidance note 2:*** Wherever possible, use standards of a different height, the work face side having a 2 m standard, and the 'outside' face having a 3 m standard. This staggers the joins, making the scaffold stronger. It also means the outside face has an extended standard of 1 m ready to receive the guardrail. ***Guidance note 3:*** Don't drive in wedges or pins at this point. These will only be driven in at step 6, after levelling and squaring.	
Step 3	While one person holds the dual standards, the second can now install the first ledger, followed by the third standard and jack combination. ***Guidance note:*** At this point the structure is free-standing; however, it is still possible in windy or otherwise unstable conditions for the standard to twist and fall – particularly if one or more screw jacks are wound out near their limit. One person should therefore continue to hold the first pair of standards.	
Step 4	Install the fourth standard and screw jack combination and position the transom (putlog) and ledger. You now have a free-standing base for the first bay. ***Guidance note:*** If you have not used the transoms and ledgers to locate the jacks, bring both components to the assembly area in readiness for this step. Don't leave your partner waiting while you walk back and forth getting the components. This increases the risk of a standard falling.	

>>

>>

Step 5	Install the putlogs and ledgers to create the first 'lift'. You will place these in at approximately 1500 mm above the base-level ledgers and transoms. ***Guidance note 1:*** The stars or 'V' clusters are set at 500 mm apart. Count three spaces and your components should be going into the fourth cluster from the bottom. ***Guidance note 2:*** With the screw jacks and sole plates taken into account, this means your first lift will be approximately 2 m from the support surface.	
Step 6	Level one of the transoms and one of the ledgers using a spirit level and the screws on the screw jack to raise or lower the standards as required. Now use your eye to sight the other ledger and transom into level. ***Guidance note:*** It is better to sight the second set of components into level, as this ensures there is no twist (known as 'wind') in the scaffold.	
Step 7	Before fixing in the wedges or pins, a few checks should now be made. In this order, check that the bay is: 1 the required distance from the structure you are scaffolding (generally less than 225 mm) 2 running parallel to the structure you are building around (assuming that is what you need it to do). Adjust the position of the standards and jacks as required 3 'square', by measuring the diagonals from standard to standard. Adjust standards and jacks as required. Now fix wedges or pins.	 
Step 8	Now continue this process for the remaining required bays. ***Guidance note 1:*** Place your ledger in position on the ground or support surface to position the screw jacks as you need them. ***Guidance note 2:*** Always have someone hold the standard until a ledger or transom is locked into position to stabilise it. ***Guidance note 3:*** Check each bay for level as you go, but also use your eye to check that ledgers are running true to the wall and in a straight line for the length of the scaffold. Also sight back through your transoms, making sure there is no twist or 'wind' developing.	

>>

>>

Step	Instructions	Photo
Step 9	Where possible, fit bracing as you proceed with the first lift of bays. See 'Bracing modular scaffolds'.	
Step 10	With the first lift of the main scaffold in place, you now need to identify the location of your ladder bay. This is usually constructed as an additional bay out from the face of the main scaffold and at one end or the other. Using the same approach as previously described, establish your screw jacks and standards, *but leave one putlog out* at this point. Check for square and level. ***Guidance note:*** If you don't leave out a putlog (or a ledger), you will find it difficult to get your access ladder inside the bay later. 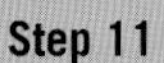	
Step 11	Position your access ladder inside the cage of the access bay and then install the last putlog. ***Guidance note:*** Ladders may soon become superseded on scaffolds, as many large companies will only allow the use of access stairs. This does not alter the standard assembly approach, however.	
Step 12	Position guardrails and midrails to the first lift for all bays using one of the options given previously (pages 178–180).	

>>

>>

Step 13	Fit scaffold planks and, where possible, toe boards to all bays by installing from below. Fit ladder access putlog to access bay and scaffold planks accordingly.	
Step 14	Having completed the first lift, repeat the procedure for all following lifts. (You are limited, of course, to only one lift more.) ***Guidance note:*** Use correct manual handling techniques for passing components up to the next level (see 'Manual handling' earlier in the chapter). Gin wheels may also be used, if necessary.	 
Step 15	Correctly position ladder, and secure to ledger, putlog, standard or transom as suitable. (This is best done with swivel couplings or ladder clips, but may be done with lashing wire or rope.) Check all wedges, pins or collars are secure. You may now access the working platform and complete any additional edge protection, such as toe boards, mesh or hoarding.	 
Step 16	Install any additional bracing required, such as around the access bay (see 'Bracing modular scaffolds' below). Once scaffold is completed, ensure any components that are not required are removed and stored away from the scaffold or removed from the site. Check all tools for any wear or damage, and then stow appropriately.	

>>

>>

Step 17	Inform the site supervisor that the scaffold is ready. Scaffold should be inspected by an independent competent person prior to being signed off, tagged and used.	

Bracing modular scaffolds

Bracing must be applied to both the face (**longitudinal bracing**) and the ends (**traverse bracing**) of the modular scaffold. Traverse bracing is to be provided at each lift at both ends of the scaffold, and is usually applied as shown in Figure 4.35. Wherever possible, braces should be attached as close as possible to the transom or putlog indicative of a lift.

Longitudinal bracing should be installed over the first and last bays of any run of scaffold and over every fourth bay in between (i.e. no more than three bays in a line may be unbraced: see Figure 4.36). Where an access bay is installed over an end bay, the face brace will go over this bay instead. The access bay is also required to have traverse bracing at least up one end. As with traverse bracing, longitudinal bracing should be attached as close as possible to ledgers indicative of each lift.

In both cases, bracing should continue to the full height of the scaffold. (Cover all lifts.)

Note: In New South Wales, reference may be made to the code of practice for this form of scaffold, which can be found at: http://www.safework.nsw.gov.au/resource-library/scaffolding/erecting,-altering-and-dismantling-scaffolding-part-1-prefabricated-steel-modular-scaffolding

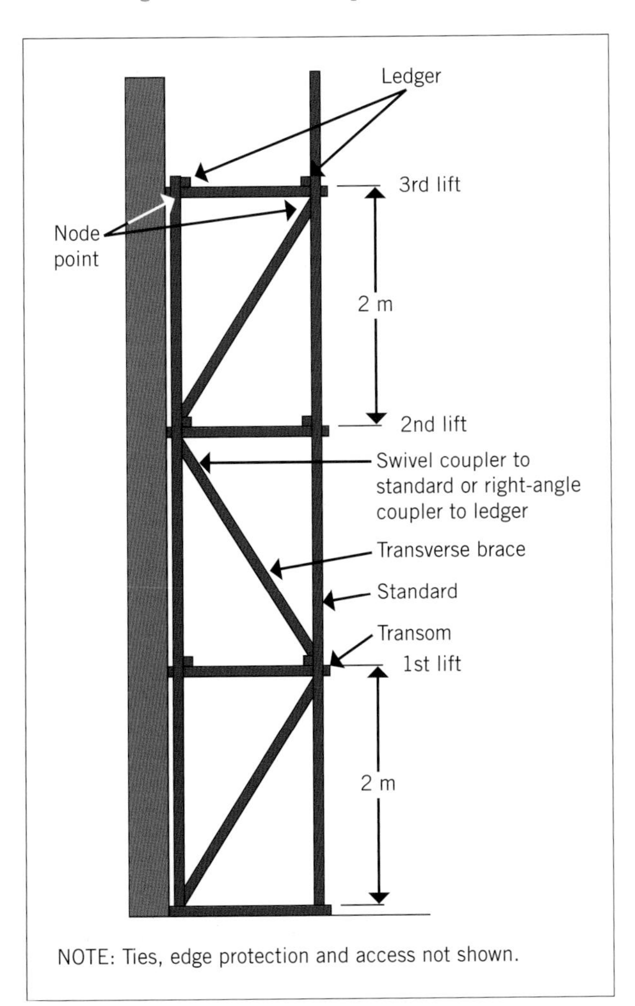

FIGURE 4.35 Traverse bracing

General scaffold dismantling procedure

Always dismantle in the reverse sequence to assembly. Keep in mind the following general precautions:

- Ensure the site supervisor and all relevant work teams are aware that the scaffold is being disassembled.
- Create a 'no go' area, or otherwise isolate the scaffold disassembly area from other workers, pedestrians, traffic, etc.
- Identify stacking and storage areas, and ensure that emergency access and services are maintained.
- Access the scaffold platforms by the ladder (or stair); never climb standards or other components.
- Pass materials down to ground (support surface) level as they are removed. Do not keep stacking them on the next convenient platform. Be constantly aware of the duty load limits of individual platforms.

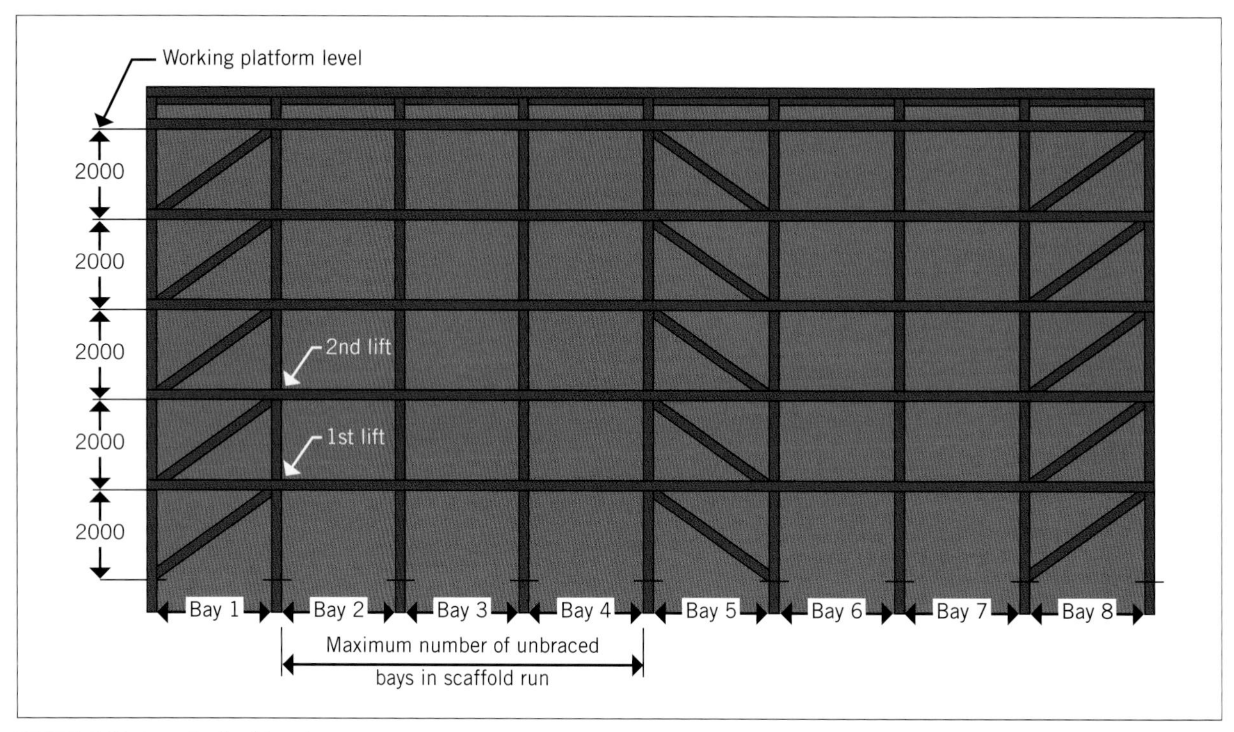

FIGURE 4.36 Longitudinal bracing

- Pass materials down; do not throw them or drop them.
- Do not leave materials lying around on platforms.
- Only work from fully planked platforms with guardrails still in place.
- Upon completion, have scaffold components removed from the site at the earliest possible moment.

Once all components have been removed, inform the site supervisor and/or principal contractor. Only with their approval may the barriers to the 'no go' area be removed and the area opened to other workers, traffic and the like.

COMPLETE WORKSHEET 3

Frame scaffold

Frame scaffolding is also known as 'unit frame', '"A" frame' or 'modular frame' scaffolding. Like modular scaffold, **frame scaffold** is a fairly common and readily available system, used to produce working platforms at heights up to 45 m (see Figure 4.37). Generally, however, frame scaffolds are of a lighter material than modular

KEY POINTS

- WHS/OHS:
  - minimum PPE is hard hat, safety footwear, HiVis vest/top, tool belt
  - manual handling – straight back, bent knees, pass – do not throw.
- Site preparation checks:
  - prior to arrival of scaffolding
  - upon delivery
  - during construction.
- Scaffolding is constructed in bays and lifts.
- Modular scaffold components may be made of steel or aluminium using one of four common jointing systems: kwik stage, cuplock, ringlock or kwik form.
- *Never* mix-and-match systems or material types in the one scaffold.
- Edge protection must be in place *before* standing on a platform where the potential fall is 2 m or greater.
- Bracing is required to both the face (longitudinal) and ends (traverse) of a scaffold.
- No more than three bays may be unbraced, and *access bays must always be braced.*
- Disassembly is the reverse of assembly – pass, never drop or throw components.

Source: Courtesy of Global Scaffold, http://www.globalscaffold.com.au

FIGURE 4.37 Frame scaffolding can be used to produce working platforms up to 45 m

systems, and so fewer working platforms per bay are allowed (typically, only one heavy-duty working platform per bay).

In addition, due to the current design of components and accessories, there are few systems available that do not expose the scaffolder to a 2 m or greater fall during assembly. Some suppliers have fashioned their own adaptors, allowing for the temporary installation of edge protection; however, these components seldom come with a certificate of compliance, nor are they part of the original design confirmation for the particular system in question.

Important note: Clause 3.10.2.1 of AS/NZS 1576.1 states categorically: 'Cross-braces on frame scaffolding *do not* satisfy the requirements for edge protection'. Even in WHS/OHS authority-approved documents outlining assembly procedures for a given frame scaffold design, this issue tends to be 'fudged' by showing the assembly of a second level of frames only. This is done without showing how to safely get edge protection established on the platform that would then need to be established above.

In due course, the approaches developed by some suppliers using additional, non-standard components may eventually be part of a confirmed design. However, as this is currently not the case, this text offers a procedure using existing frame system components only.

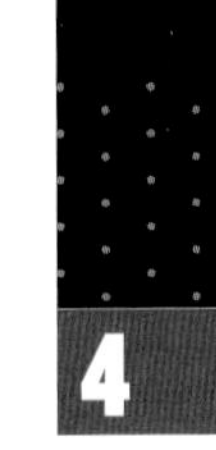

Generic assembly procedure

As with modular scaffolding, frame scaffolding is a team task requiring a minimum of two people working together at all times. In following the procedure offered, ensure that you do the following:

- Maintain good, clear communications between the scaffolding team.
- Never turn your back on a partially supported frame or brace unit unless it is clear that your partner has full control and is aware of what your next move is going to be.
- Keep tools in your belt when not using them. Never hang hammers on scaffold components, or leave any tool lying on planking.

Packing-screw jacks and pigsties

See the section 'Packing-screw jacks and pigsties' under 'Modular scaffolding'.

Options for installing edge protection

Within the generic procedures for erecting modular scaffold, three options were described. Given the commentary above on current available frame scaffolding systems, only option 1, *portable erection platforms*, is available to you.

Note: The height of individual frames in most frame scaffold systems is less than 2 m (1.8 m and 1.93 m being common). However, with the addition of sole plates, screw jacks and scaffolding planks, you must assume that entering a platform at the first lift offers a fall of greater than 2 m. Hence, full edge protection must be in place at this level before a person may stand upon it.

HOW TO

PROCEDURE FOR ASSEMBLING FRAME SCAFFOLD TO ANY HEIGHT

What follows is a sequence of steps that, if followed correctly, provides you with a safe procedure for assembling frame scaffold to any height.

Step 1	Locate the position of your first pair of screw jacks, level the surface and establish sole plates as necessary. ***Guidance note 1:*** On sloping ground it is best to start at the highest point and work down. Screw jacks at the high end should be wound down, leaving approximately 50 mm of adjustment for final levelling. ***Guidance note 2:*** Use a frame as your means of locating the jacks.	

>>

>>

Step 2	Bring two brace units and have them ready to hand. Position the first frame on the jacks. One person holds this while the two brace units are connected. This may now be left free-standing.	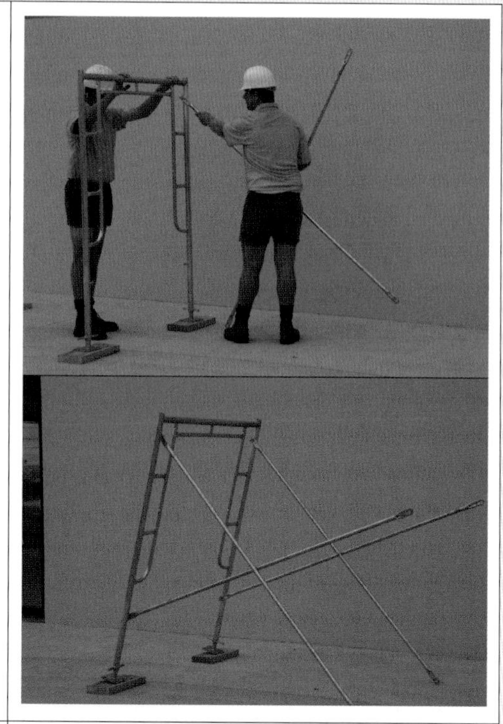
Step 3	Position your next sole plate and jack combinations based upon the bay measurement for your system (commonly 1830 mm centre to centre). ***Guidance note:*** Use a frame as your means of locating the jacks for width.	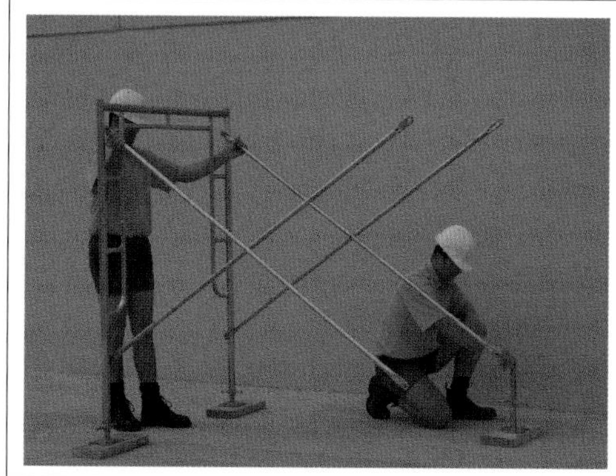
Step 4	Have one person hold the first frame and brace assembly while you stand the second frame and connect the braces. The first bay is now free-standing. ***Guidance note:*** Never let the second frame go when reaching for the brace. Stand as shown. This way you cradle the frame with your arms as you align the pins.	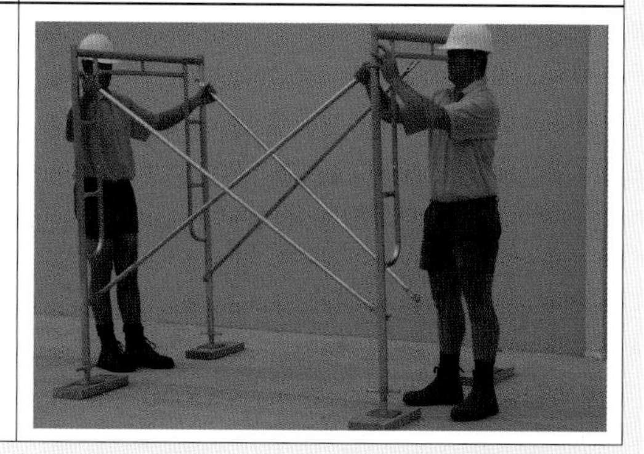

>>

>>

Step 5	Place one (only) scaffolding plank on the frames, using a spirit level and the screws on the screw jack to raise or lower the standards as required to level it. Use the spirit level to also level one of the frame heads. Now use your eye to sight the head of the other frame and use the remaining screw jack to bring it to level. ***Guidance note 1:*** It is better to sight the second frame into level, as this ensures there is no twist (known as 'wind') in the scaffold. ***Guidance note 2:*** At the same time, you can use your eye to check that the legs or standards of the two frames are aligned with each other.	
Step 6	Before going further, a few checks should now be made. In this order, check that the bay is: 1 the required distance from the structure you are scaffolding (generally less than 225 mm) 2 running parallel to the structure you are building around (assuming that is what you need it to do). Adjust the position of the frames and jacks as required 3 'square', by measuring the diagonals from frame leg to frame leg. Adjust the frames and jacks as required.	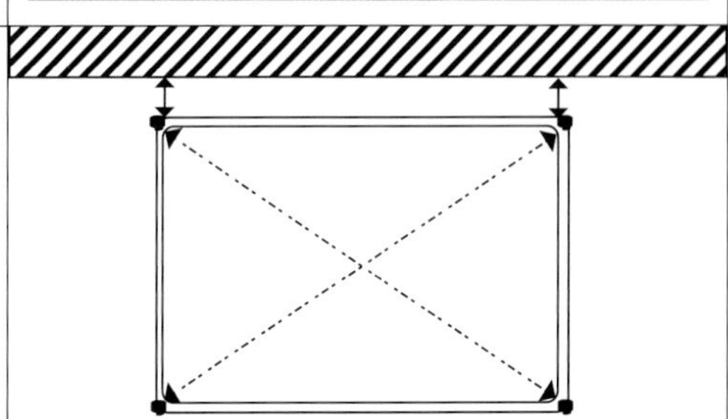
Step 7	Now continue this process for the remaining required bays. ***Guidance note 1:*** Use frames to position the screw jacks as you need them. ***Guidance note 2:*** Connect braces to the existing frame before standing the next frame. This approach actually allows for one person to stand the remaining frames on their own. The second person can follow and level each as they are installed. ***Guidance note 3:*** Continue to use your eye to check that frame heads and legs are running true to the wall, in a straight line for the length of the scaffold, and with no twist or 'wind' developing.	

>>

>>

Step	Description	
Step 8	With the first lift (without planking) of the main scaffold in place, you now need to identify the location of your ladder bay. This is usually constructed as an additional bay out from the face of the main scaffold and at one end or the other. (The photo shows the location of the ladder bay on a completed scaffold.) Using the same approach as previously described, establish your screw jacks, braces and frames. ***Guidance note:*** Depending upon the system in use, you may need to join the ladder access bay and the main scaffold together using swivel couplings (see 'Tube-and-coupler components: ties and other uses').	
Step 9	Depending upon the system being used, a ladder may be positioned inside the bay now, or a ladder access floor positioned first, then the ladder itself.	

At this point, the next stage depends upon the location of your final working platform. You will need to choose between one of the following procedures:

- procedure for where the final working platform is at the first lift
- procedure for where the final working platform is at the second lift or higher.

Note: To stay within the restraints of your 4 m height limit, you can only complete two lifts of frame scaffold.

WHERE THE FINAL WORKING PLATFORM IS AT THE FIRST LIFT

Step	Description	
Step 10	Install guardrail posts to the tops of all frame legs where edge protection will be required. Fit locking pins to protect against dislodgement.	

>>

>>

Step 11	Using step platforms of 900 mm-high folding trestle scaffold as required, position guardrails and midrails, or edge protection mesh panels (depending upon the system used). Scaffold planks are to be installed from below. *Note:* Some systems may require you to fit scaffold planks prior to edge protection mesh panels. Despite this, the handrail and midrails must be installed, prior to standing on the completed deck. ***Guidance note:*** Toe boards may be installed by accessing the deck.	
Step 12	Correctly position the ladder, and secure to the guardrail, frame, or tube-and-coupler components as required (see 'Tube-and-coupler components: ties and other uses'). ***Guidance note:*** Tube-and-coupler components may also be required to bridge any vertical gaps through which a worker may fall (see photo at step 13).	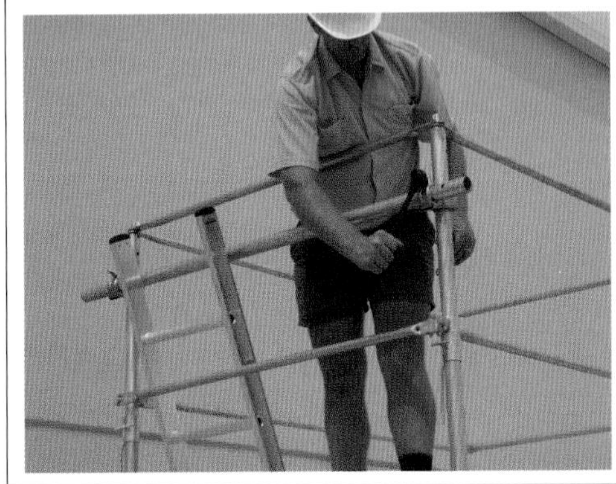
Step 13	Once scaffold is completed, ensure any components that were not required are removed and stored away from the scaffold or removed from site. Check all tools for any wear or damage, and then stow appropriately.	
Step 14	Inform the site supervisor that the scaffold is ready. Scaffold should be inspected by an independent competent person prior to sign-off and use.	

WHERE THE FINAL WORKING PLATFORM IS AT THE SECOND LIFT OR HIGHER

Step 10	Install frame-joining pins to the bottoms of the frames to be used for the second lift. Fit locking pins to protect against dislodgement.	

>>

>>

Step 11	Using step platforms or 900 mm-high folding trestle scaffold as required, position frames of the second lift on to frames of the first lift. Fit locking pins to protect against dislodgement and uplift.	
Step 12	Using step platforms or 900 mm-high folding trestle scaffold as required, fit cross-braces, midrails and guardrails or edge protection mesh panels (depending upon the system used). If going higher than two lifts you, will need to work from the platform below using step platforms. ***Guidance note 1:*** You will not be able to reach the top pins on the cross-braces. These will be done from the platform once it is installed. ***Guidance note 2:*** If going higher than two lifts, you must install the two outer planks from below first. These will act as guardrails as you install the cross-braces and rails to the lift above. ***Guidance note 3:*** Tube-and-coupler fittings may be required as guardrail for some systems. Tube must be positioned at between 0.9 m and 1.1 m above the platform. Use butt-end couplings to join lengths of tube, rather than staggering the tubing over and under each other.	
Step 13	Fit remaining scaffold planks from below. You may now access the platform. Secure top pins of cross-braces and install toe boards as required. ***Guidance note:*** You may need to relieve the coupling holding the guardrail slightly on occasion, to allow the frame pins to align with the cross-brace. Be sure to tighten couplings again afterwards.	
Step 14	The edge protection for the final platform must be installed as for 'Where the final working platform is at the first lift' above. *Note:* Scaffold planks to all bays must be installed from below. Toe boards may be installed by accessing the work platform. ***Guidance note 1:*** All edge protection must be installed by working from within the completed first lift with portable ladder platforms or the like. ***Guidance note 2:*** You must install the two outer planks from below first. These will act as guardrails as you install the cross-braces and rails to the lift above.	

>>

Step 15	Secure ladder to guardrail, frame or tube-and-coupler components as required (see 'Tube-and-coupler components: ties and other uses').	
Step 16	Once scaffold is completed, ensure that any components that were not required are removed and stored away from the scaffold or removed from site. Check all tools for any wear or damage, and then stow appropriately. Inform the site supervisor that the scaffold is ready. Scaffold should be inspected by an independent competent person prior to sign-off and use.	

Source (images): Safe High-Ts Australia Pty Ltd

Disassembly of frame scaffold

See 'General scaffold dismantling procedure' for modular scaffold (pages 185–186).

LEARNING TASK 4.4

1 **Circle 'True' or 'False'.**
When dismantling a scaffold, you should pass each piece down carefully, and nothing should be thrown to the ground.
True False

2 **When setting up the standards of modular scaffold, you should ensure that the 'V' clusters or 'stars' are aligned so that the:**
  a ledgers will be higher than the transoms/putlogs
  b ledgers will be lower than the transoms/putlogs
  c ledgers will be at the same level as the transoms/putlogs
  d 'outside' ledger will be high and the 'inside' (work face) ledger low

Bracket scaffold

Bracket scaffolds are a system of supporting frames and edge protection components designed to provide a light-duty working platform; they do so by being temporarily fixed to a structure and without recourse to direct ground support.

Bracket scaffolds are commonly erected to provide:

- working platforms for the installation of fascia and guttering
- edge protection, catch platforms, and access for roof plumbing or tiling work.

Bracket scaffolds take various forms, but may generally be described as:

- *tank bracket scaffold* – these are fixed to tanks, silos, structural steel members or the like
- *stud bracket scaffold* – used in domestic housing and fixed to the studs of wall frames (metal or timber)
- *top-plate hung bracket scaffold* – also used in domestic housing, but fixed (hung) from the top plate of the external wall.

As with all other scaffolding, bracket scaffold must comply with the AS series AS/NZS 1576 and AS 4576.

Also, as with the other scaffolds, without a Basic Scaffolding licence, you may only install bracket scaffold where a fall may be less than 4 m. In this case, you must be very careful in identifying your site restrictions; retaining walls, open trenches, even garden edging, may easily bring the fall height over the 4 m limit.

What the supplier needs to know

When purchasing or hiring a bracket scaffold system, you need to ensure the supplier or manufacturer knows what you are likely to use it for. The information you need to supply may include (but is not limited to):

- the available fixing and support structures (timber, steel, posts, stud walls)

- where on a wall or structure you intend to fix it (mid-height, from the top plate)
- the width of platform you need (will there be obstructions that would reduce the platform width to less than 450 mm: consider what the width may be after installation of guttering, for example)
- the length of platform you need (it is helpful to provide a copy of the floor or truss plan)
- the sort of work you intend doing from the platform
- the environment in which you are using the scaffold (salt air, wind, rain)
- whether you will be using the supplier's or manufacturer's scaffolding planks, or you intend to use your own (if the latter, what they are, and what distance they can span)
- whether the scaffold will act as a catch platform – in which case, it must be able to take the impact load of a falling person and/or equipment.

Without this information, the supplier cannot guarantee that the scaffold will be fit for purpose.

What you need from the supplier or manufacturer

Upon receipt of the scaffolding system from the supplier, manufacturer or hiring firm, you must check (as always) that the quantity and quality of the components are as requested. In addition, check that the supplier has provided you with documentation that clearly states:

- the make and model of the system
- that the system's design is registered with a state or national WHS/OHS authority
- that the system conforms with the relevant parts of AS/NZS 1576
- the type of planks intended to be used with the system, and that such planks have been designed and manufactured in accordance with AS/NZS 1577
- the title and edition of the written instructions provided for the safe erection, dismantling and use of the system
- the maximum load (duty rating of the system).

Generally, it is advised that you purchase or hire the scaffolding planks with the bracket scaffold system. This ensures that they are appropriate for the bracket spacings envisaged by the designers.

Edge protection to bracket scaffold

In general, the edge protection to this form of scaffold is similar to that of those outlined previously. However, as the system is less modular, and therefore more open to irregular application or positioning of components, AS/NZS 1576 stipulates the following:

- **Stanchions** (upright components rising from the brackets and to which guardrails and midrails are affixed) must be positively fixed at intervals not exceeding 3 m.
- Guardrails and midrails are to be parallel to the platform.
- Guardrails and midrails must not be more than 100 mm outside the edge of the platform.
- Guardrails must be fixed above the surface of the platform at between 0.9 m and 1.1 m.
- Midrails must be fixed approximately midway between the guardrail and the platform.
- Where the fall height from the platform exceeds 2 m, and people may be working below or around the scaffold, toe boards are to be securely fixed to the platform's open sides and ends. Toe boards should:
  - be purpose-designed components integral to the bracket scaffolding system being used, or scaffold planks placed on edge
  - extend at least 150 mm above the surface of the platform
  - be set such that no gap between the platform edge and the toe board exceeds 10 mm.
- The gap between the inside edge of the platform and the building's wall or framework must be small enough to prevent a person falling through it. If necessary, additional planks must be installed to reduce the gap.

Access and egress

A safe means of access and egress to and from bracket scaffolding must be installed before the working platform is used. This may be:

- from within the building, with the following provisos:
  - there are no gaps through which a person could slip or fall
  - there are no obstructions that have to be climbed over or crawled under
  - the means of access is readily and easily negotiable
  - when access is from an upper floor, safe access to that floor has been provided
- by portable single ladders, with the following provisos:
  - ladders are secured against movement in any direction
  - they are set up on a slope of between 4:1 and 6:1
  - they extend at least 900 mm above the surface of the platform
  - people can readily step from the ladder to the platform without climbing over or under the edge protection
  - gaps in edge protection associated with ladder access are provided with a gate, chain or other means of closing it when persons are not moving between the ladder and the platform
- by access ramps, ladder towers or temporary stair towers, with the following proviso:
  - they are constructed to comply with AS/NZS 1576.

HOW TO

PROCEDURE FOR ASSEMBLING BRACKET SCAFFOLD

Only a broadly generic installation procedure can be offered here, as there are far too many different types, and vastly different fixing systems, available for anything more specific.

As with Type 1 and 2 systems, this is a team task, requiring a minimum of two people working together at all times. You must take care to do the following:

- Maintain good, clear communications between the scaffolding team.
- Keep tools in your belt when not using them. Never hang hammers on scaffold components, or leave any tool lying on planking.
- In determining your bracket height, ensure that you position the platform such that the work may be done ergonomically – that is, to avoid overreaching, excessive bending and awkward positions.
- All installation work is to be carried out from the ground or other support surface without anyone being at risk of a fall equal to or greater than 2 m. This means you may need to install trestle scaffold, or use plasterers' folding trestles, step platforms or, failing all else, ladders, to install the system. For multi-storey construction, you may need an EWP to carry out the installation.
- Given the above, where possible, do all installation from inside the framed walls of the building.
- Ensure that platforms are nominally level, with a maximum slope in either surface direction of 3 degrees. Only if specially designed to do so, a platform may slope up to but not exceed 7 degrees along its length, but still with no more than 3 degrees across its width.
- Ensure that you have identified and are able to provide safe means of access and egress from the working platform (see 'Access and egress' on previous page).
- Check your bracket scaffold carefully, and ensure that all edge protection has been installed, before entering a platform.

Step 1	From the information provided, determine the maximum span between brackets for the scaffolding planks to be used. This will generally be not more than 1.8 m.	
Step 2	Identify the appropriate structural elements to which the brackets may be affixed. These must be no further apart than the maximum span of the supplied planks. ***Guidance note:*** It is preferred that planks be butted together rather than lapped (which presents a trip hazard). Butting is generally achieved by doubling the brackets on one support (depending upon the system used). You will need to determine at which supports double brackets are required and if each support can fit two brackets. (Some systems have a special bracket for plank junctions, which makes this step easier.)	
Step 3	Determine the height at which your brackets are to be fixed and mark this on each structural member to which a bracket will be fixed. Be sure that your location allows work to be conducted ergonomically. ***Guidance note:*** For a single run of brackets, this might be done by: • marking the two end supports only, and then using a string line or chalk line • fixing the two end brackets and then simply 'eyeing in' the others • using a laser level to position brackets, thereby eliminating the need for marking.	

>>

>>

Step 4	Fix brackets according to the approved method outlined for that particular system. ***Guidance note:*** Where possible, do this from within the building frame.	 
Step 5	Working from the ground, from inside the building, or some other means that does not expose you to a 2 m or greater fall, install edge protection (midrails and guardrails) and planking. Be sure to fit edge protection at ends of platforms. *Note:* When fitting planks: 1 the ends of planks must overhang brackets by not less than 150 mm and not more than 250 mm 2 where planks must be lapped, the upper planks should extend past their end brackets by at least 150 mm.	 
Step 6	Identify access routes and install appropriate access platforms, stairs, ladders or scaffold bays to suit.	
Step 7	Once scaffold is completed, ensure any components that were not required are removed and stored away from the scaffold or removed from the site. Check all tools for any wear or damage, and then stow appropriately. Inform the site supervisor that the scaffold is ready. Scaffold should be inspected by an independent competent person prior to sign-off and use.	

Disassembly of bracket scaffold

As always, disassembly is the reverse of installation. Take particular care to ensure that all tools and materials have been removed from platforms. Otherwise, see 'General scaffold dismantling procedure' on pages 185–186.

Trestle scaffold

Trestle scaffolds are a common form of minor scaffold that are also within the scope of AS/NZS 1576 and so must comply fully with that code in construction, installation and use. Being relatively quick and light in assembly, this method of creating a working platform is commonly used by painters, bricklayers, plasterers and, to some extent, carpenters doing fit-out and finish work.

Trestle scaffolds come in various forms, all of which are capable of creating a platform at a height greater than 2 m. As with any other situation involving potential falls of 2 m or more, full edge protection is required. The basic forms are as follows:

- **Trestle ladder scaffold** (see Figure 4.38): These consist of free-standing trestle ladders ('A' frames), scaffold planks and, where necessary, edge protection. These may incorporate stabilising arms, depending upon height.

Source: Quick Ally Scaffolding & Access Solutions (https://quickally.com.au/trestles-and-planks/safety-system)

FIGURE 4.38 Trestle ladder scaffold with handrail and outriggers

- *Frame trestle scaffold:* These consist of free-standing frames, scaffold planks and, where necessary, edge protection, and stabilising arms or **outriggers**. The frames may be 'A' or 'H' in configuration and generally fold flat upon themselves.
- *Splithead trestle scaffold:* Each 'trestle' consists of a pair of self-supporting stands that in turn support horizontal beams by means of 'splitheads' or 'U' brackets. The beams are usually scaffold planks resting on their edge, or putlogs. Scaffolding planks are then laid over to form a platform.
- *Putlog trestle scaffold:* Not a true 'trestle' as such, but rather a pair of stands, each supporting one end of a putlog, the other ends being supported by the existing structure or building. Scaffolding planks are then laid over to form a platform.

Trestle ladder and frame trestle scaffolds are generally suitable only for producing light-duty platforms for tasks such as painting and decorating, plastering or fit-out. Materials other than those in immediate use should not be stored on them. They are, however, frequently used to create platforms with potential falls of 2 m or greater.

Splithead and putlog scaffolds are more frequently used for heavier work such as brick- or block-laying. These forms of scaffold are seldom used for work over 2 m and so would not normally require edge protection unless there was a risk of falls due to the nature of the work being undertaken (such as extensive overhead work, or work with PPE such as goggles or welding masks).

What the supplier needs to know

When purchasing or hiring a trestle scaffold system, you need to ensure that the supplier or manufacturer knows what you intend to use it for. The information you need to supply may include (but is not limited to):

- the required height of your platform
- the type of work to be undertaken
- the width of platform you need (will there be obstructions that would reduce the platform width to less than 450 mm: consider what the width may be after installation of guttering, for example)
- the length of platform you need
- the number of people working from the platform at any one time
- the environment in which you are using the scaffold (salt air, wind, rain)
- whether you are using the supplier's/manufacturer's scaffolding planks, or you intend to use your own (if the latter, what they are, and what distance they can span)
- whether the scaffold will act as a catch platform – in which case, it must be able to take the impact load of a falling person and/or equipment
- whether materials (such as bricks and/or mortar) will be stored on the platform.

Without this information, the supplier cannot guarantee that the scaffold will be fit for purpose.

What you need from the supplier or manufacturer

Upon receipt of the scaffolding system from the supplier, manufacturer or hiring firm, you must check (as always) that the quantity and quality of the components are as requested. In addition, ensure that the supplier has provided you with documentation that clearly states:

- the make and model of the system
- that the system's design is registered with a state or national WHS/OHS authority
- that the system conforms with the relevant parts of AS/NZS 1576
- the type of planks intended to be used with the system, and that such planks (if supplied) have been designed and manufactured in accordance with AS/NZS 1577
- the title and edition of the written instructions provided for the safe erection, dismantling and use of the system
- the maximum load (duty rating of the system)
- the duty rating (which must also be clearly stated on the frame trestle).

Generally, it is advised that you purchase or hire the scaffolding planks with the bracket scaffold system. This ensures that they are appropriate for the bracket spacings envisaged by the designers.

Edge protection to trestle scaffold

Establishing edge protection to this form of scaffold is not easy, unless it comes as part of a purpose-designed and approved system. Designs for appropriate edge protection constructed from timber components are offered in the code of practice *Managing the Risk of Falls in Housing Construction* produced by Safe Work Australia (2018). The construction of such elements requires competence in carpentry and a firm understanding of loads. This is outside the scope of this text, and indeed the competency to which it relates. Such being the case, only purpose-designed and approved edge protection is recommended here.

Assuming the installation of purpose-designed edge protection, keep in mind these further requirements of AS/NZS 1576:

- Where the fall height from the platform exceeds 2 m, and people may be working below or around the scaffold, toe boards are to be securely fixed to the platform's open sides and ends. Toe boards should:
  - be purpose-designed components integral to the bracket scaffolding system being used, or scaffold planks placed on edge
  - extend at least 150 mm above the surface of the platform
  - be set such that no gap between the platform edge and the toe board exceeds 10 mm.
- When working next to the face of a building, the gap between the inside edge of the platform and the building's wall or framework must be small enough to prevent a person falling through it. If this is not the case, either the scaffold must be repositioned, or edge protection must be installed on this side of the platform also.

Access and egress

A safe means of access and egress to and from the platform must be installed before the working platform is used. This may be:

- the ladder of the ladder trestle scaffold, provided that:
  - materials and tools are not carried up this way except when on a tool belt
  - three points of contact are maintained at all times

  Note: Frame trestle scaffolds *must not* be climbed.
- from within the building, provided that:
  - there are no gaps through which a person could slip or fall
  - there are no obstructions that have to be climbed over or crawled under
  - the means of access is readily and easily negotiable
  - when access is from an upper floor, safe access to that floor has been provided
- by portable single ladders, provided that:
  - ladders are secured against movement in any direction
  - they are set up on a slope of between 4:1 and 6:1
  - they extend at least 900 mm above the surface of the platform
  - people can readily step from the ladder to the platform without climbing over or under the edge protection
  - gaps in edge protection associated with ladder access are provided with a gate, chain or other means of closing it when persons are not moving between the ladder and the platform
- by access ramps, ladder towers or temporary stair towers, provided that:
  - they are constructed to comply with AS/NZS 1576.

Necessary precautions when using trestle scaffold

It is essential that you take the following precautions when using a trestle scaffold:

- Know, and do not exceed, the duty load limit of the scaffold.
- Distribute loads evenly across each frame.
- As a guide, avoid more than 25 mm plank deflection under load.
- While two people may be on a two-plank scaffold, they should not:
  - stand close together in the middle of the span (maximum point load for light-duty scaffold is 100 kg: two people means a point load of approximately 200 kg)
  - attempt to move past each other to swap ends (as per the previous point, plus the platform is only 450 mm wide).
- Control movement around the scaffold by:
  - using 'no go' zones and/or dedicated travel paths
  - using spotters if the risk of inadvertent contact is high, or the work is particularly delicate
  - preventing access directly below the working platform
  - ensuring there is no work taking place above the platform.
- When working outside, monitor weather conditions and cease work in high winds or rain.
- Develop a safe approach to taking tools and materials up to the working platform by:
  - passing materials up, never throwing them
  - using a painter/plasterer's folding trestle to lift to higher platforms without excessive reaching by either those above or those below.

HOW TO

PROCEDURE FOR INSTALLATION OF TRESTLE SCAFFOLD

Installation of trestle scaffold is dependent upon the type and design of the particular system you are using. In the main, this means following the manufacturer's instructions on:

- assembly
- spacing of frames to suit supplied planks
- angle and location of braces
- installation edge protection.

The following further cautions should also be kept in mind:

- Establish a 'no go' area around where you are setting up the scaffold.
- Where locking pins are used, only use the correct pins; never use nails, screws or other metal rods. If pins are missing, you must get new ones of the correct type.
- Scaffold planks must be installed to the full width of the trestle frames or putlogs.
- Scaffold planks should be positioned by two people. Placing one end of a plank on one trestle, then trying to position the other end, is dangerous: it may lead to the collapse of the first trestle.
- Platform widths must reflect the duty load rating for which the platform is designed (see 'Working platforms and duty ratings' earlier in the chapter).
- If materials are stored on a platform, a clear width of 450 mm must be maintained for the full length of the platform.
- Scaffold planks must extend past the trestle support by at least 150 mm, but not more than 250 mm.
- Scaffold frames must be established on stable, level surfaces, or surfaces must be levelled using suitably arranged sole plates (see 'Packing-screw jacks and pigsties', page 178).
- The completed platform must be level, or deviate from level by no more than 3 degrees in either direction.
- As with all other scaffolds, ensure the scaffold and workers will not breach the required clearance distances from electrical supply lines.

Disassembly of trestle scaffold

As always, disassembly is the reverse of installation. Particular cautions to take include:

- Ensure that all tools and materials have been removed from the platform.
- Removal of planks is a two-person task. Leaving one end of a plank on a trestle may cause that trestle to collapse.

Otherwise, see 'General scaffold dismantling procedure', pages 185–186.

LEARNING TASK 4.5

1 **Circle 'True' or 'False'.**
It is acceptable for two workers to pass by each other on a two-plank trestle scaffold, provided they do so with care.
True False

2 **When using a ladder to access a trestle scaffold, it must extend:**
a 900 mm above the platform surface
b 1000 mm above the platform surface
c 900 mm above the trestle frame
d not more than 90 mm above the platform surface

Tower frame scaffold

Constructed from prefabricated components, tower frames are a specific form of scaffolding that is lightweight and mobile. Generally, **tower frame scaffold** components are made from aluminium, although steel assemblies are available. In addition, they may be fitted with lockable, height **adjustable castors**, allowing them to be moved easily over smooth, level surfaces such as concrete slabs or pathways. As most of these units are made from lightweight aluminium components, tower frame systems are generally limited to 9 m in height, unless specifically designed by the manufacturer to do otherwise.

As with other scaffolds, constructing a tower over 4 m requires one of the three scaffolder's licences outlined earlier in the chapter.

The basic components

Figure 4.39 outlines the basic components of tower frame scaffolds.

What the supplier needs to know

As with the bracket and trestle scaffold, when purchasing or hiring a tower scaffold system you need

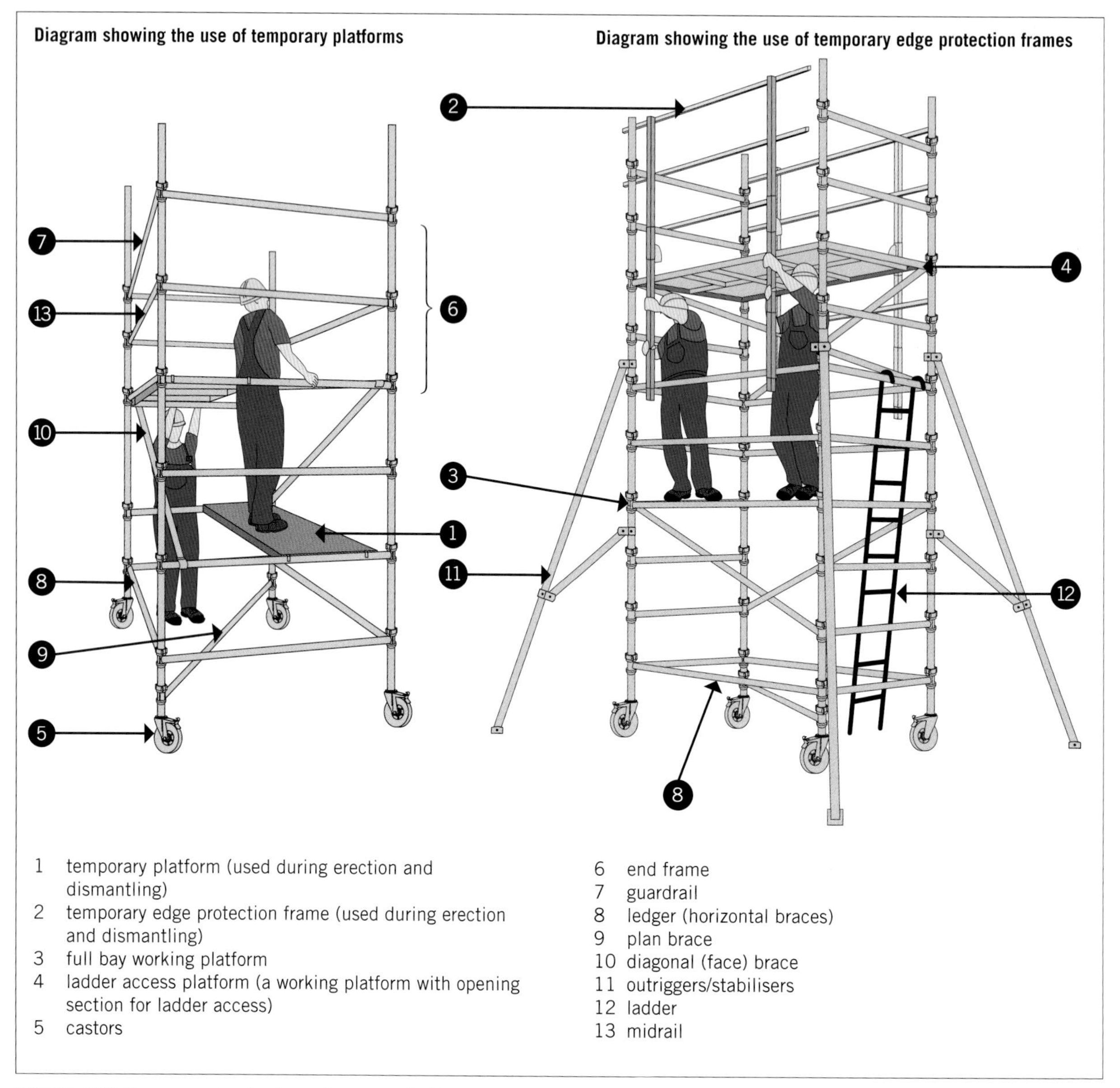

FIGURE 4.39 Components of a tower frame scaffold

to ensure that the supplier or manufacturer knows what you intend to use it for. The information you need to supply may include (but is not limited to):

- the required height of your platform
- the type of work to be undertaken
- the number of people working from the platform at any one time
- the environment in which you are using the scaffold (salt air, wind, rain)
- whether materials will be stored on the platform.

Without this information, the supplier cannot guarantee that the scaffold will be fit for purpose.

What you need from the supplier or manufacturer

Much as you would with the previous systems, upon receipt of the scaffolding components from the supplier, manufacturer or hiring firm, you should check that the quantity and quality of the components are as requested. In addition, ensure that the supplier has provided you with documentation that clearly states:

- the make and model of the system
- that the system's design is registered with a state or national WHS/OHS authority
- that the system conforms with the relevant parts of AS/NZS 1576
- the title and edition of the written instructions provided for the safe erection, dismantling and use of the system
- the maximum load (duty rating of the system)
- the duty rating (which must also be clearly stated on the platform units).

4

HOW TO

PROCEDURE FOR CONSTRUCTING TOWER FRAME SCAFFOLD

Once more, this is a team task requiring two people working together at all times. There are two distinctly different approaches offered by manufacturers of tower frame scaffold, both acceptably eliminating fall hazards greater than 2 m. In following either of the procedures offered, be sure to do the following:

- Maintain good, clear communications between you and your partner.
- Never turn your back on a partially supported frame or other component unless it is clear that your partner has full control and is aware of what your next move is going to be.
- Keep tools in your belt when not using them. Never hang hammers on scaffold components, or leave any tool lying on platform units.
- Never climb the frame of the tower; only ever use the platform units and/or the ladder(s).
- Create a 'no go' area around where the scaffold is to be assembled or disassembled.

Step 1	Read the instructions from the supplier carefully; identify and sort all components.	
Step 2	Fit **castors** to end frames; or, if using base plates, stand the frames on base plates. Fit a ledger to the bottom of each of the uprights (side vertical tubes) of the end frames (just below the bottom rail).	
Step 3	Fit **plan brace**: connects to uprights of each frame. *Note:* Some systems will require a further level of end frames to be installed prior to reaching a nominal height of 2 m. Ensure that any locking pins or other components are in place before continuing.	
Step 4	Fit face or diagonal braces as shown. *Note:* The number of face braces may differ depending upon the width, frame height or total height of the system being assembled. Check the manufacturer's instructions.	
Step 5	The base is now completed. Check: • that it is level in both directions by setting the spirit level on a frame rung, and then on a ledger • that there is no wind (twist) in the base by sighting frame to frame (rungs and uprights should be aligned).	
Step 6	Fit stabilisers.	

Source: Images by GCS

>>

>>

Now use one of the following two methods.

METHOD 1

Step 7	Install a single platform unit at approximately 1 m from the ground or support surface. This will be a temporary platform.	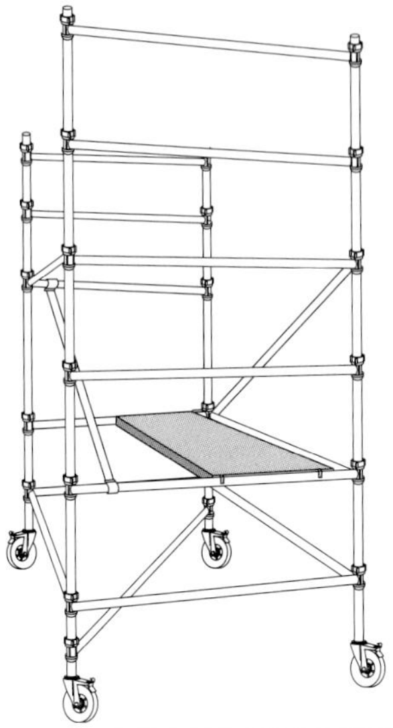
Step 8	Install a ladder access platform unit at approximately 2 m above the ground or support surface level. This unit should be on the opposite side of the frame to the temporary unit installed earlier.	
Step 9	Install a ledger opposite the ladder access platform and at the same level.	

>>

>>

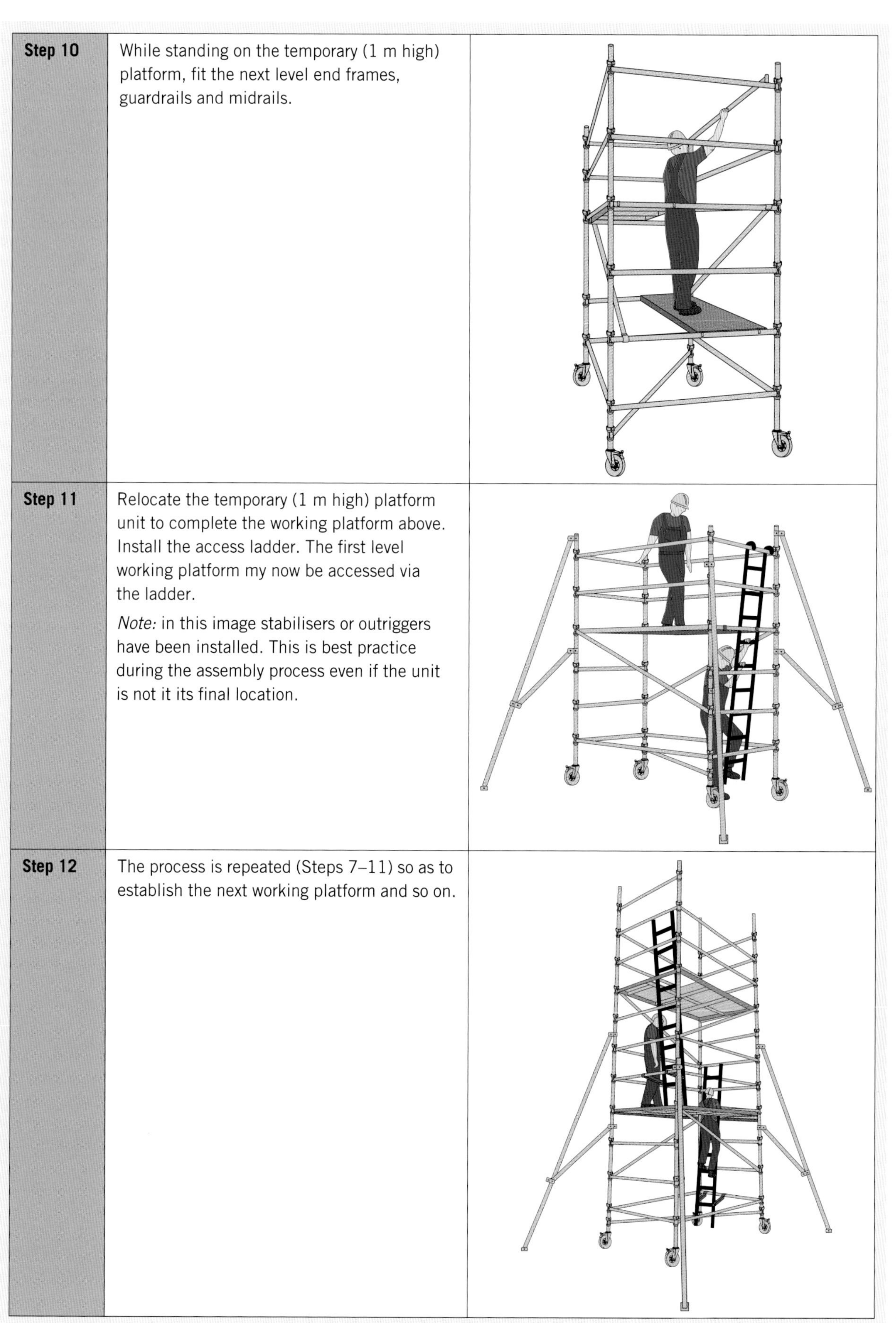

Step 10	While standing on the temporary (1 m high) platform, fit the next level end frames, guardrails and midrails.	
Step 11	Relocate the temporary (1 m high) platform unit to complete the working platform above. Install the access ladder. The first level working platform my now be accessed via the ladder. *Note:* in this image stabilisers or outriggers have been installed. This is best practice during the assembly process even if the unit is not it its final location.	
Step 12	The process is repeated (Steps 7–11) so as to establish the next working platform and so on.	

>>

>>

METHOD 2

Step 7	Install ledgers at approximately 1 m and 2 m from the ground or support surface.	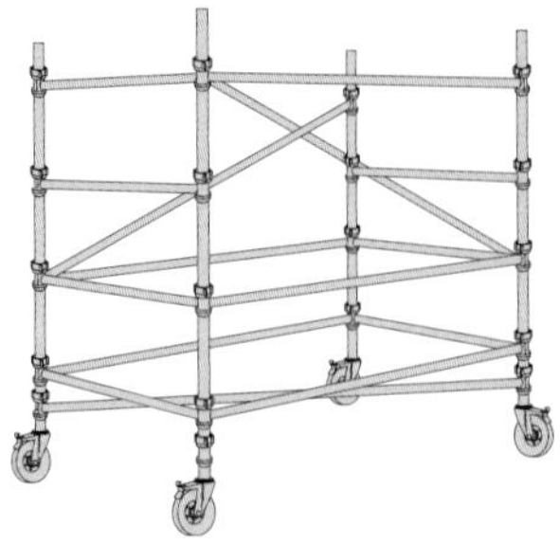
Step 8	Install additional end frames to the required height. (The top rail of the last end frame should be at least 1 m above the target working platform – that is, the first working platform will be at approximately 2 m, and the top rail of the last end frame needs to be at 3 m or greater.)	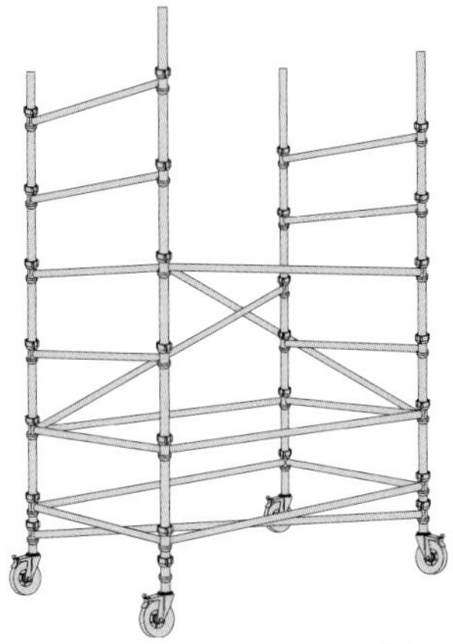
Step 9	Install platform unit, and ladder access platform units at 2 m level. Install ladder.	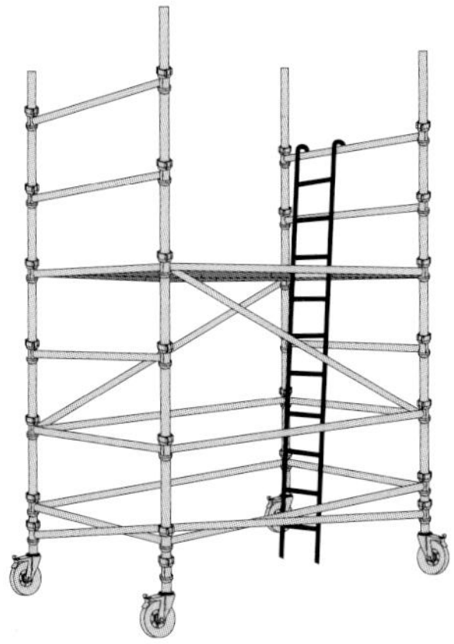
Step 10	Fit temporary edge protection as provided by, and following the directions of, the manufacturer.	

>>

>>

Step 11	Access the working platform via the ladder and install ledgers as guardrails and midrails.	

Further working platform heights may be reached by repetition of either of the above methods. Note the following:

- Where the width (shortest dimension) of the base is less than 1.2 m, the scaffold cannot exceed twice this in height without stabilisers being fitted.
- When the height reaches three times the base width (shortest dimension), stabilisers must be fitted in all cases.
- Be sure to continue to fit face (diagonal) bracing and plan bracing as required at the completion of each lift.

Disassembly of tower frames

As always, disassembly is the reverse of installation. Particular cautions to take include:

- ensure that all tools and materials have been removed from platform units
- do not remove guardrails, midrails or end frames, except by way of lower-level temporary platforms (which may need to be reinstalled as you descend)
- no components are to be dropped from the tower. All components must be passed down or lowered by ropes or other mechanical means.

Otherwise, see 'General scaffold dismantling procedure', pages 185–186.

Inspection of scaffold equipment during and after dismantling

During the dismantling process it is important to take note of any components that may have been damaged while the scaffold was in use or during the dismantling process. Any damage to scaffolding components must be reported to the supervisor so the issue can be resolved by either repair or removing the scaffold component from use.

The tools and equipment used to construct and dismantle the scaffold must also be inspected for signs of damage during the dismantling process and reported as needed.

The scaffold and the tools and equipment used in its construction should be maintained in accordance with the manufacturer's instructions to ensure they are fit for purpose and ready for use. General maintenance may require lubrication of scaffold components using specific products. These products may introduce potential hazards and should be risk assessed and the appropriate SDS reviewed for relevant safety information.

Failure to report damage or not maintaining equipment can result in serious accidents and even death.

Clean up

Once the scaffolding is dismantled it is important the area is cleaned to reduce the risk of potential tripping hazards left in the area. This includes picking up all components of the scaffolding and then storing in an appropriate location in accordance with workplace procedures or your supervisor's instructions.

The ground should be levelled if required to ensure a good working surface for other people working on-site.

Any material left over from the work completed should be reused, recycled or put into the rubbish in accordance with workplace requirements. During roofing work, off cuts of roofing material or gutter are common and left on the ground can create a serious slipping, tripping and cutting hazard.

Tube-and-coupler components: ties and other uses

As has been clearly stated elsewhere in the chapter, within the strictures of limited-height scaffolding, *you cannot construct a tube-and-coupler scaffold*. You can, however, use some of these components for minor elements of the types of scaffolds dealt with above. These components may be necessary to:

- tie a scaffold to an existing structure for additional stability
- create raking shores
- provide additional edge protection (where proprietary components will not suit)
- provide tie points for ladders or other access elements.

All tube-and-coupler components must comply with AS/NZS 1576 and AS 4576.

The components

Aside from some basic tools, there are really only two forms of components you will need to become familiar with: the tubing; and the various types of couplings used to join it with.

Tubing

Tubing is available in a variety of lengths from 1 m to 6 m. It is made from aluminium or steel (or steel alloys) and has the same range of external diameters and wall thicknesses as modular scaffold tubing.

Couplings

Tubing can be attached using several varying coupling attachments, as outlined below:

- 90-degree couplings (see Figure 4.40)
- swivel couplings
- sleeve couplings
- specialised couplings for ladders, putlogs, beams and the like.

FIGURE 4.40 Ninety-degree coupling

Tools

The main tools required have been described earlier in the chapter, these being the scaffold key, shifter and podger.

Ties and the slenderness ratio

Where the **height of a scaffold** exceeds three times its minimum base width, the scaffold must be tied to a supporting structure or fitted with raking shores (outriggers). This is often referred to as the 'slenderness ratio'. You may also need to tie your scaffold to a structure for other reasons, such as wanting increased stability due to the nature of the work being undertaken, rather than an excess in height.

When tying a scaffold into a supporting structure (Figure 4.41), each tie must:

- be rigidly connected to the supporting structure, and fixed to prevent inwards and outwards movements of the scaffold
- be connected to not less than two standards or two ledgers using 90-degree couplings

FIGURE 4.41 Tying into a building
Note: in this image the tie does not travel across two standards as required (see Figures 4.42 and 4.43). The 'check coupler', however, is in place.

- use a full-length tube without joints
- use a **check coupler** on a tie where it passes the outer lock tube.

It is best practice to stagger your ties vertically, rather than keep them in a straight line. In addition, the distance between the end of the scaffold and the first tie at any level must not exceed:

- one bay in the case of a scaffold with no return
- three bays in the case of a scaffold with a tied return
- the distance between longitudinally adjacent ties at any level shall not exceed three bays
- the vertical distance between the supporting surface and the first level of ties shall be not more than three times the least base width, subject to a maximum of 4 m
- the vertical distance between adjacent levels of ties shall not exceed 4 m, and
- the location of ties shall not obstruct clear access along the full length of any working platform or access platform.

Note: Drilled-in anchors and other methods relying on friction between the scaffold components and the supporting structure must not be used.

Figures 4.42 and 4.43 give examples of tying in a scaffold. Note the use of check couplers (also known as 'safety couplers').

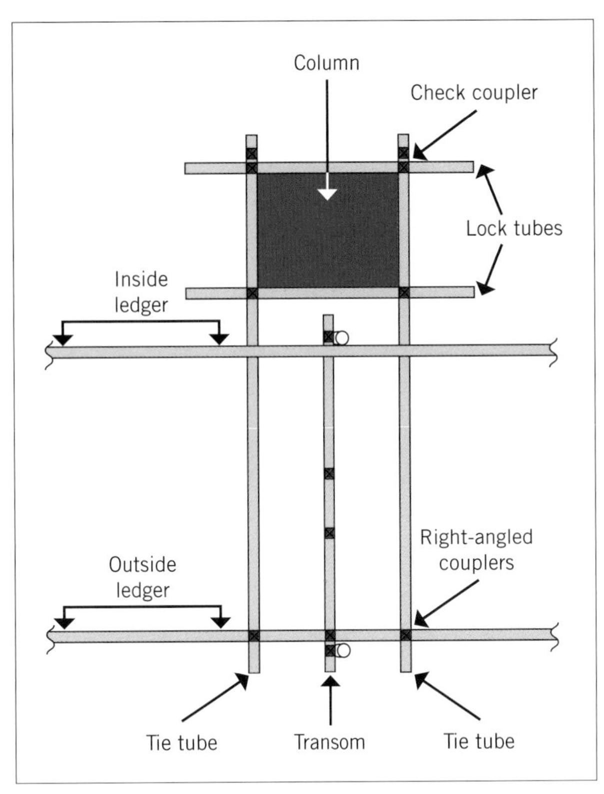

FIGURE 4.42 Tying around a column

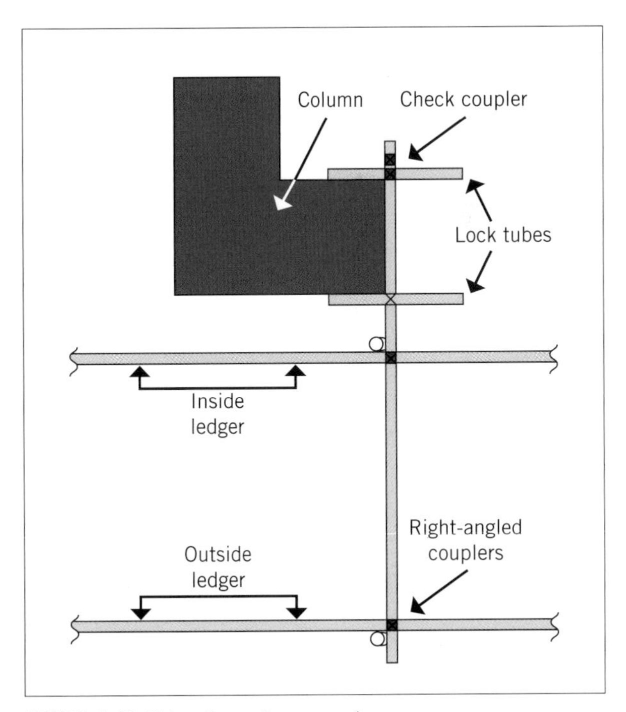

FIGURE 4.43 Tying through an opening

Raking shores

These are long sections of tubing tied high up a scaffold using swivel coupling and raking down to the ground to provide additional stability. Support-level tie-in is usually by way of a section of tube driven into the ground with a swivel coupling tying it to the raking shore.

ON-SITE

CAUTIONARY NOTE WHEN USING TIES

When tying into an existing structure, great caution must be taken to check the existing structure for structural stability and integrity. Chimneys, for example (particularly old ones), should never be used as a means of support or bracing, as they are notorious for failing under lateral loads. This also applies to old brick and block-work walls, and, indeed, similar walls that have just recently been constructed (due to uncured mortar).

KEY POINTS

- Only holders of an Advanced Scaffolding licence can construct tube-and-coupler scaffolds. However, components of tube-and-coupler may be used in limited-height scaffolding to:
  - tie scaffold to an existing structure
  - create raking shores
  - provide additional edge protection
  - tie ladders and other access elements.
- Slenderness ratio: when scaffold height exceeds 3 × base width it must be tied to a support structure or fitted with raking shores.

LEARNING TASK 4.6

1 **Circle 'True' or 'False'.**
A check coupler is a standard couple (90-degree or swivel type) that is locked on behind the main coupling on a tie tube to help prevent slippage.
True False

2 **An unlicensed scaffolder can use tube-and-coupler components to create:**
- a light-duty working platforms
- b a small suspended scaffold platform
- c raking shores
- d a personnel hoist using a gin wheel

Inspecting and maintaining the scaffold

Even though you are 'only' constructing scaffolds to a limited height, these heights are more than enough to kill or seriously injure. It is best practice, therefore, to carry out frequent inspections of, and generally maintain, your scaffold. These inspections should be:

- before the first use
- at not more than 30-day intervals
- as soon as it is safe after an occurrence that could have affected the stability or adequacy of the scaffold – for example, severe storms, earthquakes or tremors, or impact by plant, vehicles or materials
- prior to first use after alterations or repairs.

A written record should be made of the inspections, offering the following information:

- the identity number or mark, or some other means of identifying the scaffold
- pertinent design information
- the location of the scaffold
- what the scaffold is being used for
- the date and time of inspections
- comments on each inspection
- the name and signature of the person conducting the inspection.

Scaffolding labels, or 'scaf tags', with the above information should be attached to the scaffold in a visible location. All other documentation of inspections should be readily available and must be retained by the principal contractor.

What to look for

When inspecting scaffold during use, the following should be addressed:

- the scaffold has been constructed to suit the purpose for which it is being used. This is done by checking the original hazard assessment and planning documentation
- the scaffold is being used as per the original intentions and planning
- the existing inspection log and records are available and up to date
- no 'mixing and matching' of components has occurred
- appropriate edge protection is in place in zones where materials/equipment will be stacked above normal **kickboard** height
- all components are undamaged, and showing no sign of corrosion or wear
- no components are missing or have been removed or displaced by other trades
- working platforms remain open to clear, safe passage (no build-up of materials)
- footings (e.g. pigsty packing) are holding and show no sign of movement or subsidence
- vertical components have remained vertical, and horizontal components have remained horizontal
- ties are appropriately positioned and show no signs of slipping or movement
- any required alterations or repairs requested have been carried out appropriately
- hoists and gin wheels are correctly positioned and being used appropriately. Documentation regarding the installation of hoists is available (i.e. who installed them, when, their licence number and contact details, and when the hoists were last inspected).

ON-SITE

SCAFFOLD INSPECTION: LEARNING TO LOOK

Inspection of scaffolds can be done fairly rapidly, but never haphazardly. Just walking up to a scaffold with focused awareness of what 'should' be there can often tell you a lot. Start by looking at the standards and ledgers from a distance: are things straight, plumb and level? If not, look down at the base: are the sole plates sitting properly, or have they sunk into the ground; if so, why? Are workers and materials where they are supposed to be, according to the original plan? Is the work still that which the scaffold was designed for? All these things can be taken in at a glance; only then do you look for details such as location and condition of ties, corrosion or damaged components. But do not dismiss these 'small details' as minor. It is in paying attention to detail that lives can be saved: it is through inattention to detail that they are lost.

INSPECTION LOGS AND HANDOVER

When there is a change over in work personnel on a construction site, especially in management and supervisory areas, it is important that relevant information relating to scaffolding and working at heights systems is handed over to the incoming personnel.

This is usually achieved by maintaining a log book that supervisors can use to track information relating to scaffolding and other roof safety systems. This should include any information that impacts the scaffold; for example, when the last inspections were carried out or if a scaffolding needs to be altered based on the specific needs of the construction.

Workplace procedures should detail when handovers need to occur, who should be involved in the process and what information needs to be added to the log book on a day-to-day basis.

Rope, knots, gin wheels and handballing

We use ropes for a range of purposes in scaffolding, and throughout construction generally. With regard to scaffolds, you will most commonly need to use ropes to lift sections of tubing, buckets of fittings, or the materials and tools used for doing the task for which the scaffold was created in the first place. Being able to identify the appropriate rope for a given situation, and an equally appropriate knot for tying it off, is therefore an important skill. To do this, you need to know the characteristics of the various types of ropes, and the advantages and disadvantages of any given knot. Note that knots apply only to fibre ropes, not steel wire ropes.

Types of rope

AS/NZS 1576.1 allows for steel or fibre ropes to be used for purposes surrounding scaffolding. In so doing, three more standards are called up:

- AS 4142.2 Fibre ropes
- AS 3569 Steel wire ropes
- AS 2759 Steel wire rope – Use, operation and maintenance.

Fibre rope

With regard to fibre rope, the standards allow for ropes of natural or synthetic fibres to be used. In addition, two very different means of 'laying' the rope (twisting it together during manufacturing) are provided for:

- three-strand hawser-laid (Figure 4.44) – may be laid left-handed (S) or right-handed (Z)
- eight-strand plaited rope.

The three-strand hawser type is most commonly found on construction sites. It is made of either synthetic (usually polyester or polyamide [nylon]) or natural fibres, such as sisal or manila, manila being the fractionally stronger type. The construction of rope using a right-hand or 'Z' hawser lay is the most common – it is the twisting and counter-twisting of the fibres that allows the rope to remain tightly laid.

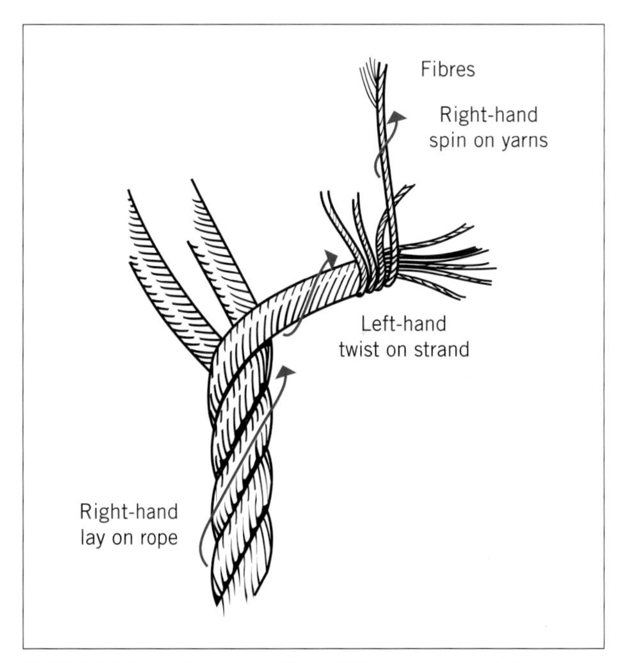

FIGURE 4.44 Basic construction of fibre ropes: right-hand hawser lay

Identifying your fibre rope

Ropes complying with the Australian Standards will have a statement to that effect on the drum or roll. Technically, they should also have an identifying coloured yarn or strand woven into them. Unfortunately, this is not often the case; and either there is no marker yarn, or it is of a colour that the manufacturer just happened to like.

Note: The coloured 'fleck' on a rope has no relevance to fibre type at all. The four main types referred to earlier may be colour coded (see Table 4.5), but care should be taken in relying upon them. If you are unsure, assume the lesser strength material.

TABLE 4.5 Identifying your rope by colour

Manila	Black
Sisal	Red
Polyamide (nylon)	Green
Polyester	Blue

Strength of fibre ropes

Of the natural fibres, manila is the strongest, with sisal being about 20 per cent weaker (although breaking strain charts treat them as one). On the other hand, sisal withstands saltwater immersion better. Of the synthetic fibres, polyamide (nylon) is the strongest but is also the only one of the various materials in that group that becomes weaker when wet, suffering a loss in strength of about 10 per cent. Despite this, it is still the preferred synthetic rope in most cases.

Tables 4.6 to 4.8 set out the breaking strains of common rope sizes, assuming that they comply with the relevant Australian Standard.

Important note: Breaking strain is *not* the safe working load (SWL) of a rope.

TABLE 4.6 Breaking strains of natural fibre rope (manila or sisal)

Diameter (mm)	Breaking strain (kN (kg))	Length of 1 kg of rope (m)	Mass per metre
10	6.22 (622)	14.70	0.068 kg
12	9.36 (936)	9.52	0.105 kg
14	12.6 (1260)	7.14	0.140 kg
16	17.7 (1770)	5.26	0.190 kg
18	21.0 (2100)	4.55	0.220 kg
20	27.9 (2790)	3.64	0.275 kg
22	33.4 (3340)	3.03	0.330 kg

Note: Conversion from kN to kg is simplified using 10 m/s^2 for gravity rather than 9.8. This is common as it gives a slight safety factor as well as being quicker.

TABLE 4.7 Breaking strains of nylon rope

Diameter (mm)	Breaking strain (kN (kg))	Length of 1 kg of rope (m)	Mass per metre of rope
10	20.4 (2040)	16.10	0.062 kg
12	29.4 (2940)	11.20	0.089 kg
14	40.2 (4020)	8.20	0.122 kg
16	52.0 (5200)	6.33	0.158 kg
18	65.7 (6570)	5.00	0.200 kg
20	81.4 (8140)	4.08	0.245 kg
22	98.0 (9800)	3.33	0.300 kg

TABLE 4.8 Breaking strains of polyester rope

Diameter (mm)	Breaking strain [kN (kg)]	Length of 1 kg of rope (m)	Mass per metre of rope
10	15.6 (1560)	13.20	0.076 kg
12	22.3 (2230)	9.09	0.110 kg
14	31.2 (3120)	6.76	0.148 kg
16	39.8 (3980)	5.13	0.195 kg
18	49.8 (4980)	4.08	0.245 kg
20	62.3 (6230)	3.30	0.303 kg
22	74.7 (7470)	2.72	0.368 kg

HOW TO

DETERMINING THE REQUIRED FIBRE ROPE SIZE FOR A GIVEN LOAD

You are required to lift 50 kg to a height of 5 m using a dry natural fibre rope. What diameter rope is required?

To do this you need to identify the appropriate rope diameter first and then do a check to see if the weight of the required length of rope (height from top of pull to the lifting surface or ground) will affect its capacity.

For example:

total load = load + rope self-load

In this instance, load equals 50 kg.

Check Table 4.6 for a breaking strain that, when divided by 10 is greater than 50 kg, with some capacity to spare.

A 10 mm diameter natural fibre rope with a breaking strain of 622 kg matches your requirements.

i.e.

$$SWL = 622 \div 10$$
$$SWL = 62.2 \text{ kg}$$

Rope self-load check

Max height of lift or length of rope = 5 m

From the same table (Table 4.6) mass per metre of 10 mm natural fibre rope = 0.068 kg

self-load of rope = rope mass per metre × required length

e.g.

$$\text{self-load of rope} = 0.068 \text{ kg/m} \times 5\text{m}$$
$$= 0.34 \text{ kg}$$
$$\text{total load} = 50 \text{ kg} + 0.34 \text{ kg}$$
$$= 50.34 \text{ kg}$$

10 mm diameter natural fibre rope with a SWL of 62.2 kg is adequate.

Note: This is still getting close to the SWL limit of this rope. It is therefore advisable to choose the next size up (12 mm or even 14 mm) so that the load being lifted lands closer to the middle of its capacity (93.6 kg and 126.0 kg respectively).

>>

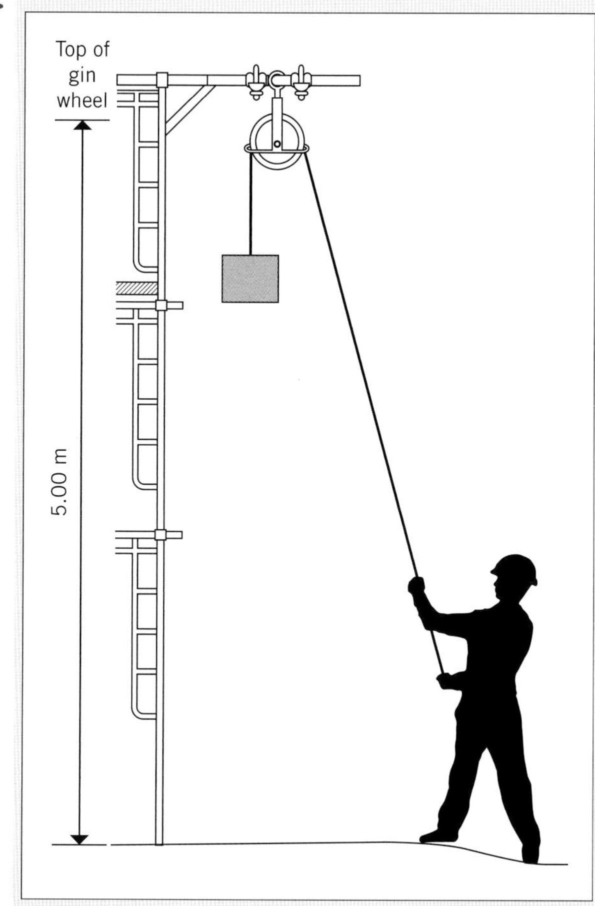

FIGURE 4.45 Gin wheel rope load

Which rope – synthetic or natural?

Although synthetic ropes are generally stronger and lighter than their natural counterparts, they are not the first choice for every occasion. The 'pros and cons' of the two fibre classes are listed below.

Synthetic

For:

- strong, less affected by water
- light, with good shock absorption
- good abrasion resistance.

Against:

- do not give any indication that they will fail (i.e. they may snap without warning)
- stretch, and so loads (when tying down) can become loose during transit
- knots are more likely to slip.

Natural

For:

- excellent knot holding
- minimal stretch after fully tensioned, and so loads tend to stay tightly held
- tend to 'give' gradually, not break suddenly.

Against:

- hold dirt and grit more readily
- heat up due to friction more readily
- abrade more readily
- is heavier when wet. Self-load per metre of rope changes dramatically from dry to wet.

Safe working load: fibre ropes

The SWL of a rope is the load that may be carried safely on that rope without *any* risk of it breaking. This is calculated for fibre ropes by dividing the breaking strain by a safety factor. AS/NZS 1576.1, clause 2.8.6.2, requires that for all fibre ropes this be a factor of 10 – that is, the SWL of all fibre ropes is 1/10th, or 10 per cent, of the breaking strain.

Using and storing fibre ropes

Fibre ropes must be carefully handled in both use and storage. Heat, cold, dust, water, moisture and ultraviolet light (UV from strong, or prolonged exposure to, sunlight) all act to degrade a rope rapidly. All ropes, whether of natural or synthetic fibres, should be stored as follows:

- hung up in a dry, well-ventilated space or in dry, ventilated lockers
- away from sunlight, flames, sources of sparks, chemicals or other agents
- laid out to dry before being stored.

HOW TO

EXAMPLE

16 mm manila rope:

SWL = 1770 ÷ 10

SWL = 177 kg

Steel wire ropes

Wire ropes are used for the supporting loads where either a fibre rope would be insufficient or a rope of smaller diameter is preferred. Made by laying high-tensile strands of wire around a core, wire rope is highly resistant to very large loadings. All wire ropes used on Australian construction sites must comply with AS 3569.

The construction of wire ropes varies significantly, and it is not within the scope of this text to cover all the varieties. Likewise, the means of designation is extremely broad, with a number of abbreviations indicating the type of core, strand, lay, grade, wire finish and overall dimension. A typical wire rope 22 mm in diameter, for example, may be designated as follows:

22 6 × 36WS-IWRC 1770 B sZ

Where:

22	=	Diameter
6 × 36WS	=	Rope construction (6 strands using a combined parallel lay)
IWRC	=	Independent Wire Rope Core
1770	=	Tensile strength of wires
B	=	Class B zinc coating
sZ	=	Right-hand lay

See AS 3569 for the full range of designators.

Safe working load: steel wire ropes

The breaking strain of steel wire ropes is given a large array of tables in AS 3569 (Tables C1–C17 of that document); otherwise, the breaking strain must be given by the manufacturer or supplier.

As with fibre ropes, however, the breaking strain is not the SWL. In the case of steel wire ropes used for lifting purposes, AS/NZS 1576.1, clause 2.8.6.1, requires that the SWL be not greater than 1/6th of the breaking strain.

FIGURE 4.46 Thimbles are fitted inside loops and then clamped with a ferrel

HOW TO

EXAMPLE

SWL = Breaking strain ÷ 6

So, for our 22 mm 6-strand rope described earlier, which the manufacturer guarantees has a minimum breaking strain of 305 kN (30 500 kg), the SWL would equal:

$$\begin{aligned} SWL &= 30\,500 \div 6 \\ &= 5083 \text{ kg} \end{aligned}$$

Note: An alternative approach approved in AS 4576 for use when the breaking strain is not known is as follows:

Working load limit (WLL) in kg = $7.5 \times D^2$

Where D = the diameter of the wire rope in mm

For example:

$$\begin{aligned} WLL &= 7.5 \times 22^2 \\ &= 3630 \text{ kg} \end{aligned}$$

(a notable reduction in allowable weight)

FIGURE 4.47 Three clamps must be used to secure the wire from slipping, as shown

Terminations

The ends of wire rope fray easily, and, as it is difficult to tie into knots, wire rope is not easily connected to other materials or equipment. The frayed ends also make it a dangerous 'sharp' on a job. Various means of overcoming these issues have been devised, though for the most part these are some form of a loop. Irrespective of the method used, all terminations must comply with AS 2759. The more common approaches are described below.

- *Thimbles* (Figure 4.46): Fitted inside loops to limit bends becoming too tight and stressing the cable. Cable is then clamped, or 'swaged', with a ferrel (as in the figure).
- *Wire rope clamps* (Figure 4.47): A wire rope clamp, effectively a miniature 'U' bolt, is used to fix the loose end of a loop of wire back on itself. Three clamps must be used to secure the wire from slipping. These must be fitted as shown in the figure.
- *Swaged terminations* (Figure 4.48): Swaging (pronounced 'swayging') is a means of fixing a socket, ferrule or other item to a wire end by effectively squashing the component into the wire. All sorts of terminations are possible with this method, including studs, turnbuckles, ferrules and the like.

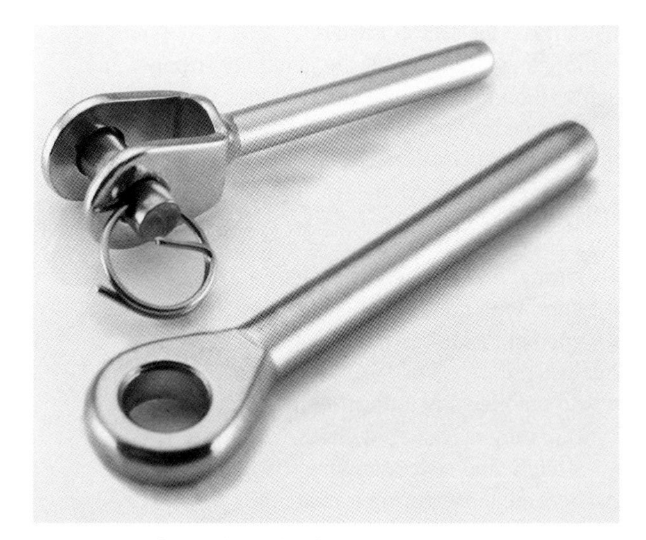

FIGURE 4.48 Swaged terminations

- *Wedge sockets*: These are useful when it is likely that a fitting will be regularly replaced or repositioned. As the load increases on these fittings, the wedge is pulled tighter.

Bends and hitches

Aside from knowing a particular rope's breaking strain and how to calculate its SWL, a sound knowledge of bends and hitches, how to tie them, and where to use them is important. The importance of knowing 'true' knots (which is how you might commonly have referred to a bend or hitch) is not just that they are less likely to let go unexpectedly; they are less likely to lock up and not be able to be undone, and less likely to create a point load that will break the rope. A basic description of the most common knots used in scaffolding (and construction generally) follows. Note that bends and hitches are used with fibre ropes, not steel wire ones.

Warning: Different types of fibre ropes can respond differently to the same bend or hitch. In addition, if you learn knots from books or the internet only, you run the risk of tying them incorrectly without realising it. Learn how to tie from an experienced instructor, and become familiar with how effective they are on different fibre types through practice and experience.

The most common bends and hitches used in scaffolding are:

- *Alpine butterfly* (Figure 4.49): Used when a secure loop is needed in the middle of a piece of rope. Excellent mid-line rigging knot and handles multi-directional loading well

FIGURE 4.49 Alpine butterfly

- *half hitch* (Figure 4.50): A basic knot that is used singly or in multiples to reduce the likely slippage of other knots
- *clove hitch* (Figure 4.50): Often used for tying your rope to a rail, tube or bar before tying down a load. Not to be used for lifting. Best completed with a half hitch
- *round turn and two half hitches* (Figure 4.51): Used for securing a rope or line to posts, rails, tubes and the like. This knot has far superior holding power than the clove hitch

FIGURE 4.50 Clove hitch with half hitch

FIGURE 4.51 Round turn and two half hitches

- *bowline* (Figure 4.52): Can be used as a secure loop for hooking on to posts and the like. A particularly strong knot that will not allow the loop to become tight around the object it is hooked over
- *figure of 8* (Figure 4.53): A 'stopper' knot, used to stop a rope from coming out of a pulley or guide. Also used as a backup on the tail of other knots
- *timber hitch* (Figure 4.54): A quick, temporary hold to components for the purpose of dragging or lifting when used in conjunction with a half hitch further up the piece
- *rolling hitch* (Figure 4.55): Can be used to tie on to a pole, tube or other round components for lifting when the load pulls the rope in parallel with the component. Offers superior and more permanent holding than the timber hitch. With a slight variation, it may also be tied on to another rope, or be used for passing the rope around a post or peg and then tying the rope back on to itself

FIGURE 4.52 Bowline

FIGURE 4.53 Figure of 8

- *reef knot* (Figure 4.56): Also called a 'square knot'. Used to join two ropes of equal size together. Good for parcels and first aid, but should be avoided when critical loads are considered. When over-tensioned, it can either lock up or slip
- *sheet bend* (Figure 4.57): Excellent all-round knot for joining ropes together. When joining ropes of significantly different thickness, the knot can be improved by adding extra turns, as shown in Figure 4.57.

For an animated breakdown on how to tie these (and many other) bends and hitches, go to http://www.animatedknots.com.

Gin wheels

Note: Gin wheels may only be installed by someone with a Basic Scaffolding licence

FIGURE 4.54 Timber hitch

FIGURE 4.55 Rolling hitch with additional hitch

FIGURE 4.56 Reef knot

Gin wheels are used as a means of getting materials and equipment up to a working platform without having to manually carry them up stairs and ladders. Effectively, they are a simple pulley attached to a

FIGURE 4.57 Sheet bend (doubled)

protrusion from the scaffold (see Figure 4.58) through which a rope is passed (fibre if operated by hand, steel wire if mechanised).

FIGURE 4.58 Gin wheel

Safe installation and operation

When installing a gin wheel, it must be positioned:

- not more than 600 mm along an unbraced cantilevered piece of scaffolding tube
- as close as practicable to the brace on the scaffolding tube.

When operating a gin wheel:

- not more than 50 kg may be lifted by the gin wheel irrespective of the SWL of the wheel itself or the rope used
- rope should be of sufficient length to allow the operator to stand well clear of the load being lifted
- the area below the gin wheel and around the operator should be a 'no go' zone to other workers while in operation
- only gin wheels with rope guides shall be used
- the minimum diameter fibre rope to be used is 16 mm.

Handballing

This is the name given to a manual method of passing scaffolding components up to the next lift. One person stands on the ground and passes the component up to the next person standing above, and so on up the scaffold (Figure 4.59). Handballing is an approved means of getting individual components up a small number of lifts, but must be conducted with great care. The main points to keep in mind are:

- maintain a 'no go' zone around the area where handballing is to take place. In determining this space, consider the length of the components being passed up
- do not overreach either in passing or receiving the components
- workers must keep their bodies behind the guardrails at all times
- after about three or four lifts, handballing is an inefficient means of getting components aloft; consider alternatives

FIGURE 4.59 Handballing in action

- having passed a component up, the person beneath should stand well clear until the item is secure within the scaffold aloft. They should also keep their eye on the component for its entire passage
- workers standing on intermediate working platforms, having passed the item on, should likewise stand well back from the railing until 'Clear' is called out from the worker at the final lift.

KEY POINTS

- Steel or fibre ropes may be used in scaffolding for lifting, tying down or pulling up components.
- Ropes must comply with safe working loads (SWL):
  - for fibre ropes it is 1/10th of breaking strain
  - for steel wire ropes it is 1/6th of breaking strain.
- Gin wheels may be used for getting materials and equipment up to working platforms, provided:
  - no more than 50 kg is lifted at any one time
  - a 'no go' area surrounds the gin wheel operation.
- Handballing involves teams of workers stationed at different lifts to pass and receive components during construction or disassembly of the scaffold.

LEARNING TASK 4.7

1 **Circle 'True' or 'False'.**
The value in tying 'true' knots is that they reduce point load pressures on the ropes and they are easy to undo.
True False

2 **The SWL of a 5 mm 7 × 7 strand stainless steel wire rope with a breaking strain of 1770 kg is:**
a 177.0 kg
b 295.0 kg
c 17.70 kg
d 29.50 kg

Consolidation: creating the SWMS

Earlier in this chapter you were told that for all scaffolding work a SWMS must be created. At that time you were not in a position to finalise such documents because you simply did not know enough about scaffolding, or the implications of any choices or decisions you might make. Being now somewhat more informed, and – it is hoped – with some practical experience gained as well, you should be able to consolidate your knowledge with the generation of the appropriate documentation.

This process has been outlined in each of the three previous chapters, but is summarised again below:

1 *Identify* the hazard or potential hazard.
2 *Assess* the level of risk and determine the priority using a risk management matrix. (See Appendix 1 in Chapter 2, Tables A.2, for an example risk management matrix.)
3 *Control* the risk using the hierarchy of control shown in Figure 3.18 in Chapter 3.
4 *Review and evaluate* the *risk and the control measures* on a continual basis.

The first part you did in your researches on site context and task requirements. The second part is making the necessary decisions on what scaffold to use and how to use it. This involves breaking down the task and determining the level of risk and countermeasures involved.

In creating your SWMS, be sure to include input from as many of the relevant stakeholders as possible. All workers involved in the task must agree to and sign the finished document.

Having documented an agreed process, follow it! If, upon reflection or in the doing of the task, it becomes obvious that the process needs changing, do so – but again, all those involved need to be advised of the changes and once again sign on to it.

LEARNING TASK 4.8

1 **Circle 'True' or 'False'.**
In the hierarchy of control, elimination is always your preferred option where possible.
True False

2 **When developing your SWMS document, you should always consider the position of all stakeholders. In scaffolding these may be:**
a other workers
b components of raking shores
c pedestrian traffic
d both 'a' and 'c'

GREEN TIP

When planning for the erection of scaffold, ensure that the weather during the time that the scaffold will be erected is not going to create additional hazards for the workers. Also ensure the scaffold damages the ground as little as is necessary for a safe scaffold.

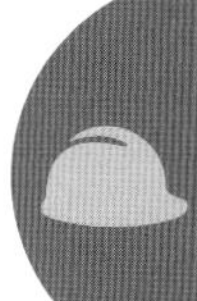

Always make sure that scaffolding is tagged either suitable for use or unsuitable for use so all workers in the vicinity of the scaffold know the status of that scaffold. Scaffolding of any height can be very dangerous and many injuries have occurred from lack of care.

COMPLETE WORKSHEET 4

SUMMARY

Good scaffolding is safe, efficient and effective. The scaffold should be able to be positioned and removed quickly without risk of injury to either the scaffolder or those around them. In addition, it should give a stable working platform for the required tasks to be achieved. To aid you in developing the skills and knowledge to create such structures, the chapter covered a range of factors:

- The introduction outlined the various scaffolding types, where they might be used, and who may legally construct them.
- The importance of carefully planning the job through a five-step process consolidated into a SWMS document was recommended. In addition, you were advised to pay particular attention to minimum safe distances from power lines and what must be done when it is necessary to reduce those distances.
- Choosing the right scaffold is one of the outcomes of the planning process. This choice reflects your determinations with regard to WHS/OHS, the location and context, and, of course, the purpose of the structure in the first place.
- This 'purpose' directs attention to the working platforms and duty ratings of such platforms, what loads are required to be carried, and for how long, etc. These loads generally reflect worker and material movements, which in turn suggest the types of access and edge protection required, such as handrails, kickboards and sometimes hoarding.
- The assembly and bracing of five types of scaffolding have been outlined fully in the chapter to ensure that this is conducted in a safe manner, limiting potential falls to less than 2 m.
- Tube-and-coupler components have also been discussed, but within the limits of the qualification, meaning that such elements may only be used as ties, raking shores, additional edge protection and improving the integrity of access points.
- Once constructed, a scaffold must undergo continuous inspection regimes. Failure to do this can mean the scaffold may be unsafe due to any number of reasons, from storm damage to deliberate vandalism. It is equally important, throughout this process, to identify any part of a scaffold that has been altered by untrained/unauthorised workers trying to improve their access to a particular element of the building under construction.
- The second to last section details the handling of ropes (steel and fibre), particularly important knots, plus gin wheels and handballing techniques for getting materials and scaffolding components to higher levels. The importance of working within SWLs, and how to calculate these, was also addressed within this section.
- The chapter closed with suggestions about how you might consolidate all of this information and knowledge into a well-informed SWMS document.

Ultimately, your job is to not only produce the right scaffold without risk of injury, but also provide one that offers a well-founded sense of security. That is, to not only to make a scaffold look and be sound, it should also feel sound. The best scaffolds are those where a person feels no difference when stepping between the solid structure and your piece of handiwork. If you take someone up who is not keen on heights and you see fear in their eyes, it is safe to say you have not done your job.

REFERENCES AND FURTHER READING

Text

Australian Safety and Compensation Council (2006), *National Standard for Licensing Persons Performing High Risk Work*, Australian Safety and Compensation Council, Canberra, **https://www.safeworkaustralia.gov.au/system/files/documents/1702/nationalstandard_licensingpersonsperforminghighriskwork_2006_pdf.pdf**.

Safe Work Australia (2018) *Managing the Risk of Falls in Housing Construction: Code of Practice*, Commonwealth of Australia, Canberra, **https://www.safeworkaustralia.gov.au/system/files/documents/1810/model-cop-preventing-falls-in-housing-construction.pdf**

Safe Work Australia (2019a), *Model Work Health and Safety Act*, Commonwealth of Australia, Canberra, **https://www.safeworkaustralia.gov.au/doc/model-work-health-and-safety-act**

Safe Work Australia (2019b), *Model Work Health and Safety Regulations*, Commonwealth of Australia, Canberra, **https://www.safeworkaustralia.gov.au/doc/model-work-health-and-safety-regulations**

Relevant Australian Standards
AS/NZS 1576.1:2019 Scaffolding – Part 1: General requirements
AS/NZS 1576.2:2016 Scaffolding – Part 2: Couplers and accessories
AS 1576.3:2015 Scaffolding – Part 3: Prefabricated and tube-and-coupler scaffolding
AS/NZS 1576.4:2013 Scaffolding – Part 4: Suspended scaffolding
AS/NZS 1576.5:2021 Scaffolding – Part 5: Prefabricated splitheads and trestles
AS 1576.6:2020 Scaffolding – Metal tube-and-coupler scaffolding – Deemed to comply with AS/NZS 1576.1
AS 4576:2020 Guidelines for scaffolding
AS/NZS 1577:2018 Scaffolding planks
AS/NZS 1801:1997 Occupational protective helmets
AS/NZS 1163:2016 Cold-formed structural steel hollow sections
AS/NZS 1664.1:1997 Aluminium structures – Limit state design (reconfirmed 2020)
AS/NZS 1664.2:1997 Aluminium structures – Allowable stress design – Commentary (Reconfirmed 2020)
AS/NZS 1866:1997 Aluminium and aluminium alloys – Extruded rod, bar, solid and hollow shapes (Reconfirmed 2020)
AS/NZS 1734:1997 1734 Aluminium and aluminium alloys – Flat sheet, coiled sheet and plate
AS 1892.1:2018 Portable ladders – Part 1: Performance and geometric requirements
AS 1892.2:1992 Portable ladders – Part 2: Timber (Reconfirmed 2022)
AS 1892.5:2020 Portable ladders – Part 5: Selection, safe use and care
AS 4142.2:1993 Fibre ropes (Current status is 'Withdrawn' – but it still may be referrenced as nothing supersedes it)
AS 3569:2010 Steel wire ropes – Product specification
AS 2759:2004 Steel wire rope – Use, operation and maintenance

GET IT RIGHT

Source: mrscaffold.com.au

FIGURE 4.60 Scaffolding handrail

In Figure 4.60, a scaffolding handrail component at right is limiting your access to the balcony face and under-surface. Based upon what you have learnt in this chapter, identify what actions must be taken, and by whom, to temporarily remove this component or otherwise alter the scaffold to allow the access you need.

___

___

___

WORKSHEET 1

To be completed by teachers	
Student competent	☐
Student not yet competent	☐

Student name: ______________________________

Enrolment year: ______________________________

Class code: ______________________________

Competency name/Number: ______________________________

CPCCCM2008: Erect and dismantle restricted height scaffolding – specifically, the elements 'Plan and prepare' and 'Erect scaffolding'.

Task: Answer the following questions.

1 At completion of your training under the competency unit CPCCCM2008: Erect and dismantle restricted height scaffolding, you may construct scaffolds to what height?

a 2 metre

b 4 metres

c 6 metres

d 10 metres

2 Circle 'True' or 'False'.

At completion of your training under the competency unit CPCCCM2008: Erect and dismantle restricted height scaffolding, you are only eligible to apply for a Basic Scaffolding licence. Further training is required for Intermediate and advanced licences.	True	False

3 When would you be required to hold a Basic Scaffolding licence?

____

____

____

4 List the five types of scaffold discussed in the chapter.

1 ______________________________

2 ______________________________

3 ______________________________

4 ______________________________

5 ______________________________

5 What is a 'suspended' scaffold, and who may construct it?

What: ______________________

____

____

____

Who: ______________________

____

6 List six of the nine points that WHS/OHS authorities require with regard to scaffolding.

1 ______________________

____

2 ______________________

____

3 ______________________

____

4 ______________________

____

5 ______________________

____

6 ______________________

____

7 Which Australian Standards suite applies specifically to scaffolding?

____

____

____

____

____

8 Planning a scaffolding job, particularly hazard assessment, revolves around inspections and good communications. Two key elements of the approach are discussions with stakeholders and ensuring public and/or worker access. List the three other key elements.

1 ______________________

2 ______________________

3 ______________________

9 List seven stakeholders (aside from other workers) with whom you may need to consult prior to selecting and constructing a scaffold.

1 ______________________________

2 ______________________________

3 ______________________________

4 ______________________________

5 ______________________________

6 ______________________________

7 ______________________________

10 What is considered the best solution to 'crowd control' around a scaffold? List seven ways by which this may be achieved.

What:

____

Means of establishing:

1 ______________________________

2 ______________________________

3 ______________________________

4 ______________________________

5 ______________________________

6 ______________________________

7 ______________________________

11 What is the role of a spotter when constructing scaffolding?

____

____

12 Frequently, it is necessary to obtain a permit before constructing a scaffold. This may be from the local council, water authority or even the police. List four more authorities or agencies from which a permit may be needed.

1 ______________________________

2 ______________________________

3 ______________________________

4 ______________________________

13 Identified hazards may fall into three categories: support surface, scaffold erection area, and overhead and other elevated hazards. List four possible hazards for each of these categories.

Support surface:

1 ______

2 ______

3 ______

4 ______

Erection area:

1 ______

2 ______

3 ______

4 ______

Overhead and elevated:

1 ______

2 ______

3 ______

4 ______

14 AS 4576 provides for specific distances from electrical power lines. Which regulations override these distances? What is the minimum distance that any part of an unshielded scaffold can be from a 66 kV line (street power lines on poles) in your state or territory?

Overriding regulation: ______

______

Minimum distance: ______

______

WORKSHEET 2

Student name: ______________________________

Enrolment year: ______________________________

Class code: ______________________________

Competency name/Number: ______________________________

To be completed by teachers	
Student competent	☐
Student not yet competent	☐

CPCCCM2008: Erect and dismantle restricted height scaffolding – specifically, the elements 'Plan and prepare' and 'Erect scaffolding'.

Task: Answer the following questions.

1 When choosing the right scaffold, it is suggested you should consider your choice from three perspectives: WHS/OHS, location and context, and purpose. List four considerations for each of these perspectives.

WHS/OHS:

1 ______________________________

2 ______________________________

3 ______________________________

4 ______________________________

Location and context:

1 ______________________________

2 ______________________________

3 ______________________________

4 ______________________________

Purpose:

1 ______________________________

2 ______________________________

3 ______________________________

4 ______________________________

2 Working platforms are rated for a particular duty. List the three duty ratings, including their maximum loads, point loads and minimum widths.

1 ______________________________

2 ______________________________

3 ______________________________

3 What Australian Standard applies to scaffolding planks?

____

4 Describe or define the location of the following components.

Standard: ____

____

____

Putlog: ____

____

____

Transom: ____

____

____

Scaffold plank: ____

____

____

Toe board: ____

____

____

Midrail: ____

____

____

Guardrail: ____

____

____

5 Circle 'True' or 'False'.

The minimum width for a scaffolding plank is 220 mm (nominally 225). True False

6 The minimum thickness of an LVL scaffolding plank is:

a 38 mm
b 35 mm
c 32 mm
d 40 mm

7 List five defects that would prohibit you from using a timber scaffolding plank.

1 ______________________________

2 ______________________________

3 ______________________________

4 ______________________________

5 ______________________________

8 What is the minimum clear passage that must be maintained on any working platform (irrespective of its duty rating)?

____

____

9 Materials and equipment cannot be stored on a light-duty platform because:

a Being frequently only 450 mm wide, a clear passage of 450 mm cannot be maintained

b They are not strong enough to carry much more than a person

c There is no toe board to prevent tools or materials from being knocked off

d Both (a) and (b)

10 List five requirements that working platforms must comply with.

1 ______________________________

2 ______________________________

3 ______________________________

4 ______________________________

5 ______________________________

11 What is meant by 'edge protection' with regard to scaffolding and working platforms? When must it be installed?

Edge protection: ______________________________

____

____

When installed: ______________________________

12 Circle 'True' or 'False'.

Wire or fibre ropes can be used as access closures but must not be used as guard rails. True False

13 The work face or building side of the platform does not need edge protection, provided that four criteria are met. List the four criteria below.

1 ______

2 ______

3 ______

4 ______

14 List six means by which access to the work platforms of scaffolds is generally provided.

1 ______

2 ______

3 ______

4 ______

5 ______

6 ______

15 Circle 'True' or 'False'.

Access ramps must only be installed by holders of Intermediate (or higher) licences	True	False

16 Who may install a personnel hoist?

______

______

17 An access ramp with a slope greater than 7 degrees or 1:8 must be cleated. What does cleating mean?

______

______

18 What is a 'barrow ramp' and how is it different from a standard sloping access ramp?

______

______

WORKSHEET 3

Student name: ______________________________

Enrolment year: ______________________________

Class code: ______________________________

Competency name/Number: ______________________________

To be completed by teachers	
Student competent	☐
Student not yet competent	☐

CPCCCM2008: Erect and dismantle restricted height scaffolding – specifically, the elements 'Plan and prepare' and 'Erect scaffolding'.

Task: Answer the following questions.

1 List four items of PPE that must be worn when assembling scaffolding.

1 ______________________________

2 ______________________________

3 ______________________________

4 ______________________________

2 Individual fall-arrest systems (IFAS) and travel restraint systems:

a Are not considered appropriate for general scaffold construction because of entanglement and no clear fall zone

b Must never be used on any type of scaffolding project due to the dangers of entanglement and lack of clear fall zone

c Must be worn at all times with a shortened lanyard to reduce entanglement issues

d Cannot be worn due to the need to wear a scaffolder's tool belt

3 What are the seven basics of manual handling as they apply to scaffolding?

1 ______________________________

2 ______________________________

3 ______________________________

4 ______________________________

5 ______________________________

6 ______________________________

7 ______________________________

4 If the scaffold is incomplete and you have to leave the site, you must:

a Install barriers to block access to work platforms, including signs warning that the scaffold is incomplete and entry is prohibited

b Fence off the area (support surface) around the scaffold to prevent access, including signage warning that the scaffold is incomplete and entry is prohibited

c Ensure the scaffold is at a point whereby it is secure against collapse due to climatic imposed loads such as wind, rain, snow and the like

d All of the above

5 List the basic tools used in scaffold assembly.

____

____

____

____

____

____

____

6 When preparing the site prior to receiving the scaffolding components, what must be done?

____

____

____

____

____

____

____

____

____

7 Upon receipt of scaffolding components at the site, what five things should you do?

1 ______________________________

2 ______________________________

3 ______________________________

4 ______________________________

5 ______________________________

8 What is the purpose of sole plates under screw jacks?

____

____

____

____

____

____

9 What is meant by 'pigsty' packing and why is it done?

____

____

10 What is the minimum length of screw jack that should be within a standard?

a 50 mm

b 100 mm

c 150 mm

d 200 mm

11 Why must you not mix scaffolding components of different materials (such as aluminium and steel) in the one scaffold?

____

____

12 A supplier provides you with components that are of the same type of scaffold and materials (all steel or all aluminium), but they are a mix of brands or manufacturers. What must the supplier also provide to you?

____

____

13 In the section 'Generic assembly procedure' (p. 185) you are advised to always keep your tools in your belt when not in use, and never leave tools on planks or scaffold components. Why do you believe this is important advice?

____

____

14 What is meant by 'bays' and 'lifts' in scaffolding?

Bay: ______________________________

Lift: ______________________________

WORKSHEET 4

To be completed by teachers	
Student competent	☐
Student not yet competent	☐

Student name: ______________________________

Enrolment year: ______________________________

Class code: ______________________________

Competency name/Number: ______________________________

CPCCCM2008: Erect and dismantle restricted height scaffolding – specifically, the elements 'Inspect, repair and alter scaffolding', 'Dismantle scaffolding' and 'Clean up'.

Task: Answer the following questions.

1 Circle 'True' or 'False'.

You may construct a tube-and-coupler scaffold without a licence provided you do not exceed 4 m in height to the working platform. True False

2 Aside from the slenderness ratio, what other factors might require you to tie the scaffold to the main structure or use outriggers (raking shores)?

____

3 A 'check coupler' is:

a Used to restrict or prevent slippage of a load-bearing coupler along the tube
b A special form of coupler that can swivel to notched angles
c A coupler temporarily positioned to check a proposed platform height
d Used to determine the location of the next putlog on tube and coupling scaffold

4 Circle 'True' or 'False'.

Drilled in anchors (such as Dynabolts and loxins and other friction fixings) must never be used as a means of tying a scaffold to a structure. True False

5 The slenderness ratio requires a scaffold be tied to a support structure or be fitted with raking shores when:

a The width is three times greater than the height
b The height is three times greater than the width
c The height of the upper platform is over 4 m
d It is constructed as a tower scaffold

6 Circle 'True' or 'False'.

Scaffolds must be inspected before first use, at 30-day intervals, after any incident and prior to use after repair or alteration. True False

7 What seven pieces of information must be included in the written record of inspections?

1 ______________________________

2 ______________________________

3 ______________________________

4 ______________________________

5 ______________________________

6 ______________________________

7 ______________________________

8 Who must retain the inspection documentation?

____

____

____

9 You have inspected a scaffolding that has been constructed for a 2-storey house so the plumbers can install the fascia and gutter. The fascia and gutter have been installed and the plumbers are now preparing to install the roof cladding. They currently have a step ladder on the scaffold to access the roof because the work platform is too low. Explain the steps you should take to ensure the scaffold is fit for purpose and the code of practice that should be referenced.

____

____

____

____

10 What is a 'scaf tag'?

____

11 When inspecting scaffolding, what are 10 things you should address?

1 ______________________________

2 ______________________________

3 ______________________________

4 ______________________________

5 ______________________________

6 ______________________________

7 ______________________________

8 ______________________________

9 ______________________________

10 ______________________________

12 When inspecting scaffold, what do you look for first, and when?

What: ______________________________

____

When: ______________________________

____

13 Alterations or repairs are found to be required after you have inspected the scaffold. When would the next inspection be due?

____

____

14 What is a gin wheel used for, and what is the maximum load that may be lifted by one?

Use: ______________________________

____

____

Maximum load: ______________________________

15 Handballing is:

a A technique for strengthening and loosening up the joints in the hands prior to assembling or disassembling a scaffold

b A means of rolling heavy scaffolding components up to the scaffold in preparation for lifting

c A manual handling method of passing scaffolding components up to the next lift or work platform

d The expression given to lifting materials and scaffolding components using ropes and a pulley (gin wheel) mounted on the top level of the scaffold

16 Circle 'True' or 'False'.

Handballing becomes an inefficient means of getting components up a scaffold once it goes beyond three or four lifts.	True	False

17 Describe when it would be best to use the following knots.

Round turn and two half hitches: ______

Bowline: ______

Sheet bend: ______

Figure of 8: ______

18 When dismantling trestle scaffold, one particular caution you should observe is that all materials and tools have been removed from the work platform. What is the other caution?

______

19 When dismantling tower scaffold, what are three particular cautions that should be observed?

1 ______

2 ______

3 ______

WORKSHEET 5

To be completed by teachers	
Student competent	☐
Student not yet competent	☐

Student name: ______________________

Enrolment year: ______________________

Class code: ______________________

Competency name/Number: ______________________

Task: Answer the following questions.

1 You have been asked to organise some scaffolding to install new guttering on a two-storey house. The scaffold provider has asked what type of duty rating the scaffold needs to be: light, medium or heavy.

Based on the following information, calculate the duty rating required. Refer to Table 4.1 and Table 4.2 in the chapter to assist you. Show your calculations in the space provided.

- The scaffolding will have two (2) people, weighing 80 kg and 105 kg respectively, using one (1) platform.
- The gutter and other materials on the platform weigh 10 kg.
- The tools and equipment needed for the installation weigh 30 kg.
- The width of each bay does not matter as the scaffold will be installed around the perimeter of the building.

____

____

____

____

____

____

2 You have just received a phone call and one of the plumbers that was going to use the scaffolding has been called to another job, so the 2nd year apprentice is coming over to assist in the installation. Based on this change, reassess whether a lighter duty rating be used instead.

- The scaffolding will have two (2) people, weighing 65 kg and 105 kg respectively, using one (1) platform.
- The gutter and other materials on the platform weigh 10 kg.
- The tools and equipment needed for the installation weigh 30 kg.
- The width of each bay does not matter as the scaffold will be installed around the perimeter of the building.

GLOSSARY

A

access platform A platform that gives access to and from places of work to persons, materials and equipment.

adaptor A variety of fittings that allow a PA tool to be subtly converted, allowing it to be used for different purposes or with different materials.

adjustable base plate A base plate with an adjustable leg.

adjustable castor A castor incorporating a height-adjusting device that fits inside the standard or vertical member of a scaffold.

administrative controls Documents that outline a required or preferred course of action, required skills and/or supervisory tasks to be adhered to. In short, administrative controls state who, when, where and how a work activity is to be undertaken.

anchorage point A secure point for attaching a lanyard or other element of a travel restraint or fall-arrest system. Anchorages are designed and rated for their particular use.

authorised person The manufacturer, their nominee or a qualified gunsmith.

B

backfill The earth put into a hole, trench or other excavation to fill it. Always suspect as a supporting surface.

barrel That part of the PA tool down which the fastener is propelled (also the piston in low-velocity tools).

barricade An easy to see barrier that closes or blocks off an area, road or path.

base material The material into which a fastener will ultimately be driven and held.

base plate A plate that is able to distribute the load from a load-bearing member to a supporting structure.

bay A section of scaffolding enclosed by four adjacent standards, or the equivalent space in a single-pole scaffold.

birdcage scaffold An independent scaffold consisting of more than two rows of standards connected by ledgers and transoms. Generally used to produce platforms that are greater than one bay in width.

bleed-down valve (also known as an *emergency descent valve*) A valve in the hydraulic system that drains oil in the pistons or cylinders back to the reservoir, and in so doing allows the EWP to be lowered when the pump or motor has failed.

brace (Chapter 1) A diagonal component used to stabilise or prevent undue movement of frames, scaffolds, walls or the like.

brace (Chapter 4) A member fixed diagonally to two or more members of a scaffold to provide rigidity to the scaffold.

bracket scaffold A scaffold that has a platform carried on frames attached to or supported by a permanent or temporary construction.

C

cartridge A gunpowder-filled brass cylinder that, when strck by a firing pin, will explode. Used in PA tools as the means of propulsion for fasteners.

castor A swivelling wheel attached to the lower end of a standard, for the purpose of supporting and moving a scaffold.

catch platform A platform, attached to a scaffold, to contain falling debris. The structure must be designed to withstand the impact load of a falling person, persons and/or materials.

charge *See* cartridge.

check coupler A right-angle coupler or swivel coupler that is fixed hard against a load-bearing coupler, to restrict or prevent slippage of that coupler along the tube.

closed platform A platform that is capable of being a working platform but is temporarily closed to any loading or access by persons in accordance with the installation design.

code of practice A guide developed by state and/or national work safety authorities offering advice on hazard identification and safe work practices.

competent person A person whose experience, knowledge, skills and qualifications provide them with the ability to perform and/or supervise a specific task or group of tasks.

coupler A fitting that joins two tubes.

cradle The portion of a suspended scaffold that incorporates a suspended platform.

cycling The action of manipulating a PA tool to bring the piston, charge and firing pin into position ready for firing.

D

deadman pedal/switch A switch or pedal that must be held on at all times by the operator for the controls to be activated. Can be a foot pedal or hand switch.

decal A sticker on an EWP that provides procedural, warning or specification information.

direct acting A type of PA tool in which the force of the exploding cartridge is transferred directly to the fastener. Also known as 'high-velocity tools'.

duty of care A burden of responsibility, or legal obligation, to have thought for the safety of anyone who may be affected by your actions or, indeed, failure to act.

E

elevating work platform (EWP) (also known as *elevated work platform*) A mobile machine (device) that is intended to move persons, tools and material to working positions and consists of at least a work platform with controls, an extending structure and a chassis, but does not include mast-climbing work platforms.

emergency stop button A button which, when pressed, stops all operation of the machine in the case of an emergency. Generally, it is a large red button.

energy absorber A device fitted between the lanyard and the safety harness designed to limit the fall-arrest force applied to the body.

explosive-powered tool (EPT) *See* PA tool.

extension ladder A non-self-supporting portable ladder that is adjustable in length and consists of two or more sections with guides or brackets to permit length adjustment.

F

fall-arrest system A system designed to stop you from falling more than a predetermined distance, and to slow you down towards the end of that distance.

fasteners Drive pins or threaded studs designed to be fired into a hard surface by means of a PA tool.

firing pin A small, spring-loaded pin that, when released by the trigger, strikes the cartridge and causes it to explode.

frame scaffold A scaffold assembled from prefabricated frames, braces and accessories.

G

gantry A structure, constructed from structural steel, scaffolding or structural timber, that is primarily intended to support a protection deck or portable buildings such as amenity sheds.

gradability The maximum angle of the ground over which the machine may be mobilised with the EWP in the lowered position.

guardrail A structural member to prevent persons from falling off any platform, walkway, stairway or landing.

H

hazard With regard to falls from a height, any situation where there is potential for someone to fall from one level to another.

height of a scaffold The vertical distance from the supporting structure to the highest working platform of the scaffold.

I

independent scaffold A scaffold consisting of two or more rows of standards connected together longitudinally and transversely.

indirect acting A type of PA tool in which the force of the exploding cartridge is transferred to a piston, which then 'pushes' the fastener into the material. Also known as 'low-velocity tools'.

individual fall-arrest system (IFAS) A system designed to stop a person from falling to the ground. Consists of a harness and lanyard assembly.

industrial rope access system May be part of a travel restraint system or work positioning system. This line (generally steel wire rope) is used for gaining access to, and/or working at, a work face.

inertia reel (also known as *self-retracting lanyard* or *fall-arrest block*) A mechanical device that arrests a fall by locking up a lanyard when it is subjected to a sudden movement much as a car seat belt does. Yet, like a car seat belt, it still allows the lanyard to travel to its full length if drawn out steadily.

J

job safety analysis (JSA) A form of administrative control used to identify and document hazards and/or risks applicable to a particular scope of work. Generally used in conjunction with a SWMS or SWP.

K

kickboard *See* toe board.

kickplate *See* toe board.

L

ladder An appliance on which a person may ascend or descend, consisting of two stiles joined at regular intervals by cross-pieces (e.g. cleats, rungs, steps, treads).

landing A level area providing access to a stairway or ladder, or located at an intermediate level in a system of stairways or ladders.

lanyard (or lanyard assembly) A length of line (generally webbing) that connects between the harness worn by the worker, and an appropriate anchorage on the building, structure or EWP. A lanyard assembly consists of the lanyard and a personal energy absorber. The lanyard, and or lanyard assembly, should be as short as reasonably practicable, with a working length of not more than 2 m.

ledger A horizontal structural member that longitudinally spans a scaffold (runs the long side of a rectangular scaffold bay).

licensed The person holds a relevant qualification from an appropriate body.

lift The vertical distance from the supporting surface to the lowest ledger of a scaffold or level at which a platform can be constructed. Also, the vertical distance between adjacent ledgers or levels of a scaffold at which a platform can be constructed.

longitudinal bracing Bracing in a vertical plane on the face of a scaffold.

M

magazine An adaptor that allows the loading of multiple fasteners to a semi-automatic or fully automatic tool.

maximum operating angle (also known as *chassis inclination, slope sensor setting, operating incline maximum*) The maximum safe ground angle the machine is designed to be elevated on as specified by the manufacturer.

melding The subtle fusion of the surface of a PA fastener with the base material into which it has been forced.

midrail A member fixed parallel to and above a platform, between the guardrail and the platform.

misfire When a charge or cartridge fails to explode upon impact of the firing pin.

mobile scaffold An independent free-standing scaffold that is mounted on castors.

modular scaffold A scaffold assembled from prefabricated individual components, braces and accessories.

N

'no go' area A clearly defined area into which people, other than those required to complete the task(s) involved, are restricted from entering.

O

operator The person controlling the EWP.

outrigger (Chapter 4) A framed component that increases the effective base dimensions of a tower and is attached to the vertical load-bearing members.

outrigger/stabiliser (Chapter 3) Mechanical legs and/or jacks used to support the EWP and improve stability by levelling and increasing the footprint of the machine on the ground.

P

PA tool Powder-actuated tool – a tool designed to fire fasteners into hard surfaces such as steel or concrete. They do so by means of a cartridge or charge similar in form to a .22 calibre firearm.

passive fall prevention device Material or equipment, or a combination of material and equipment, that is designed to prevent a person from falling, and which, after initial installation, requires ongoing inspections to ensure its integrity but does not require ongoing adjustment, alteration or operation to perform its function. Examples include scaffolding and perimeter guardrailing.

personal protective equipment (PPE) The last line of defence against a hazard – for example, eye or hearing protection, hard hat, gloves, safety harness, etc.

pigsty packing Lengths of hard wood positioned to create a stack or bed for the outrigger footplate to sit on to distribute the weight of the EWP over a larger surface area on the ground.

piston The part of an indirect-acting PA tool that pushes the fastener down the barrel. It is captive within the PA tool and cannot exit the barrel.

plan brace A brace in the horizontal plane that is attached to vertical load-bearing members.

platform An elevated surface – that is, that part of an EWP upon which an operator stands when being elevated and working at heights; or the raised surface of a scaffold upon which workers will stand. Also carries tools and equipment.

platform height The vertical distance from the surface upon which the EWP is supported to the floor of the platform (basket) at its maximum height.

platform reach The horizontal distance from the axis of rotation of the boom to the outer edge of the platform when fully extended.

podger hammer A steel tool used for the locking and releasing of typical modular scaffolding fixing devices.

power take-off (PTO) A small gearbox attached to a truck's main gearbox. When engaged, it drives the hydraulic pump that powers the boom.

prefabricated scaffold A scaffold assembled from prefabricated components and manufactured so that the geometry of the scaffolding is predetermined.

putlog A horizontal structural member, spanning between ledgers or between a ledger and an adjacent wall, that is intended to support a platform.

R

raking shore An inclined tube fixed to a scaffold to keep the scaffold stable.

reasonably practicable What could or should be done when the likelihood of risk, severity of possible injury, costs, and existing knowledge and skill are considered together.

return A part of a scaffold set up around the corner of a building or structure.

risk The likelihood of an injury or some harm (such as illness) occurring coupled with the possible level of injury or harm that may result.

road plate Flat steel or thick-ply plate used to improve the distribution of the load from an outrigger.

S

safe work method statement (SWMS) A form of administrative control used to identify and document control measures and safe work practices for a given scope of work.

safe work procedure (SWP) A form of administrative control used to identify and document control measures and safe work practices for a given scope of work.

safe working load (SWL) The total weight (in kilograms) that can be on a work platform, or lifted by a cable, rope, chain or EWP. The combined weight of operators, materials, tools and equipment is not to exceed the SWL rating for any given machine.

scaffold A temporary structure specifically erected to support one or more access platforms or working platforms.

scaffold key A scaffold spanner.

scaffold plank A decking component, other than a prefabricated platform, that is able to be used in the construction of a platform supported by a scaffold.

scaffold spanner A box or tube-type wrench with a swing-over handle that has been purpose-designed for the tightening and releasing of couplers.

scaffolder A person engaged in erecting, altering or dismantling scaffolding.

side force The maximum allowed sideways force (push or pull) that can be applied to the platform.

single ladder A non-self-supporting portable ladder of fixed length and consisting of one section.

single-pole scaffold A scaffold consisting of a single row of standards that are connected together by ledgers and putlogs fixed to ledgers and built into the wall of a building or structure.

sole plate A member used to distribute a load through a base plate to the ground or other supporting structure.

spalled The damaged surface of concrete or brickwork brought about by the impact of previous fasteners or some other force. Spalled surfaces contain fractures and loose material which cannot be fired into with PA tools.

spotter Someone who acts as your eyes and ears, looking around to make sure no one has entered a 'no go' area, or that you or others are otherwise in a dangerous position prior to you using specialised equipment.

stanchion A vertical member used to support a guardrail, a mesh panel or similar.

standard A vertical structural member that transmits a load to a supporting structure.

step ladder A self-supporting portable ladder of fixed length having flat steps or treads and hinged back legs.

stile (also known as a side rail) A member in a ladder that supports rungs, steps or treads.

supporting structure A structure, structural member or foundation that supports a scaffold.

T

tare weight The unladen weight of a vehicle.

tie A member or assembly of members used to tie (link or hold) a scaffold to a supporting structure.

toe board A scaffold plank or purpose-designed component fixed at the edge of a platform to prevent material falling from the platform.

tower frame scaffold A prefabricated scaffold consisting of fabricated units and only able to be erected in the form of a tower.

transom A horizontal structural member transversely spanning an independent scaffold between standards (runs the short side of a rectangular scaffold bay).

travel restraint system (also known as *work positioning system*) A system incorporating a harness or belt that is attached to one or more lanyards. The lanyards are attached in turn to a static line or anchorage point. By this means, the travelling range of a worker is restricted, preventing them from reaching a position on a structure from which they could fall.

traverse bracing Bracing in a plane that is vertical and at right angles to the building or structure.

trestle ladder scaffold A scaffold consisting of trestle ladders supporting scaffold planks.

trestle scaffold A scaffold consisting of trestles and planks.

tube-and-coupler scaffold A scaffold of which the standards, ledgers, braces and ties are circular tubes that are joined together by means of purpose-designed couplers.

W

wind rating The maximum wind speed an EWP is designed to operate in when elevated.

work positioning system (also known as *travel restraint system*) A system incorporating a harness or belt that is attached to one or more lanyards. The lanyards are attached in turn to a static line or anchorage point. By this means the travelling range of a worker is restricted, preventing them from reaching a position on a structure from which they could fall.

working height The height the operator can reach and work when the EWP is at its maximum extended height.

working load limit The maximum working load that may be applied to any component or system, under general conditions of use.

working platform A platform that is intended to support persons, materials and equipment.

INDEX

Note: Page numbers in bold represent defined terms.

F

G

H

I

J

K

L

M

N

O

P

U

V

W